DIRECTOR'S CHOICE

MUSEUM OF THE HISTORY OF SCIENCE OXFORD

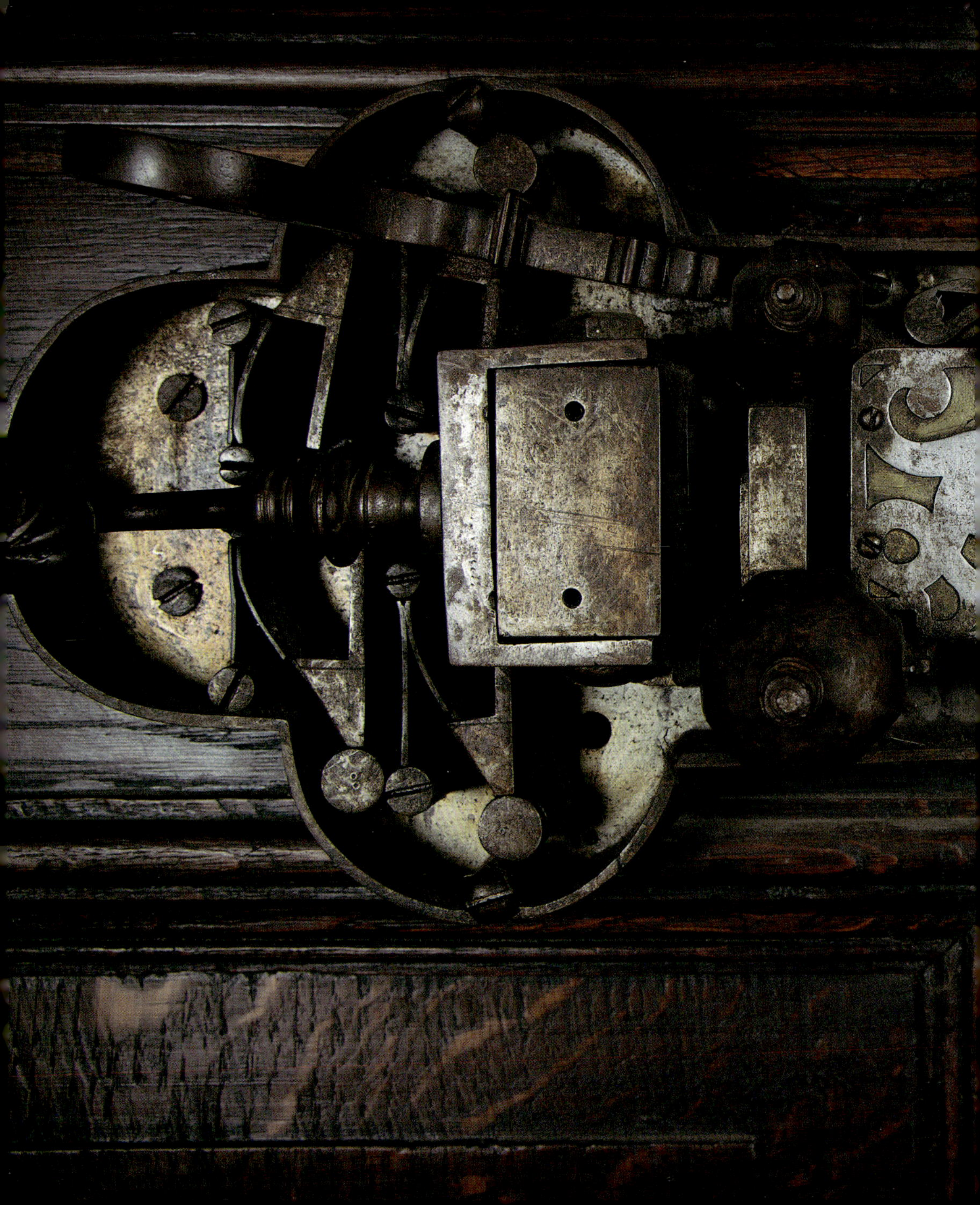

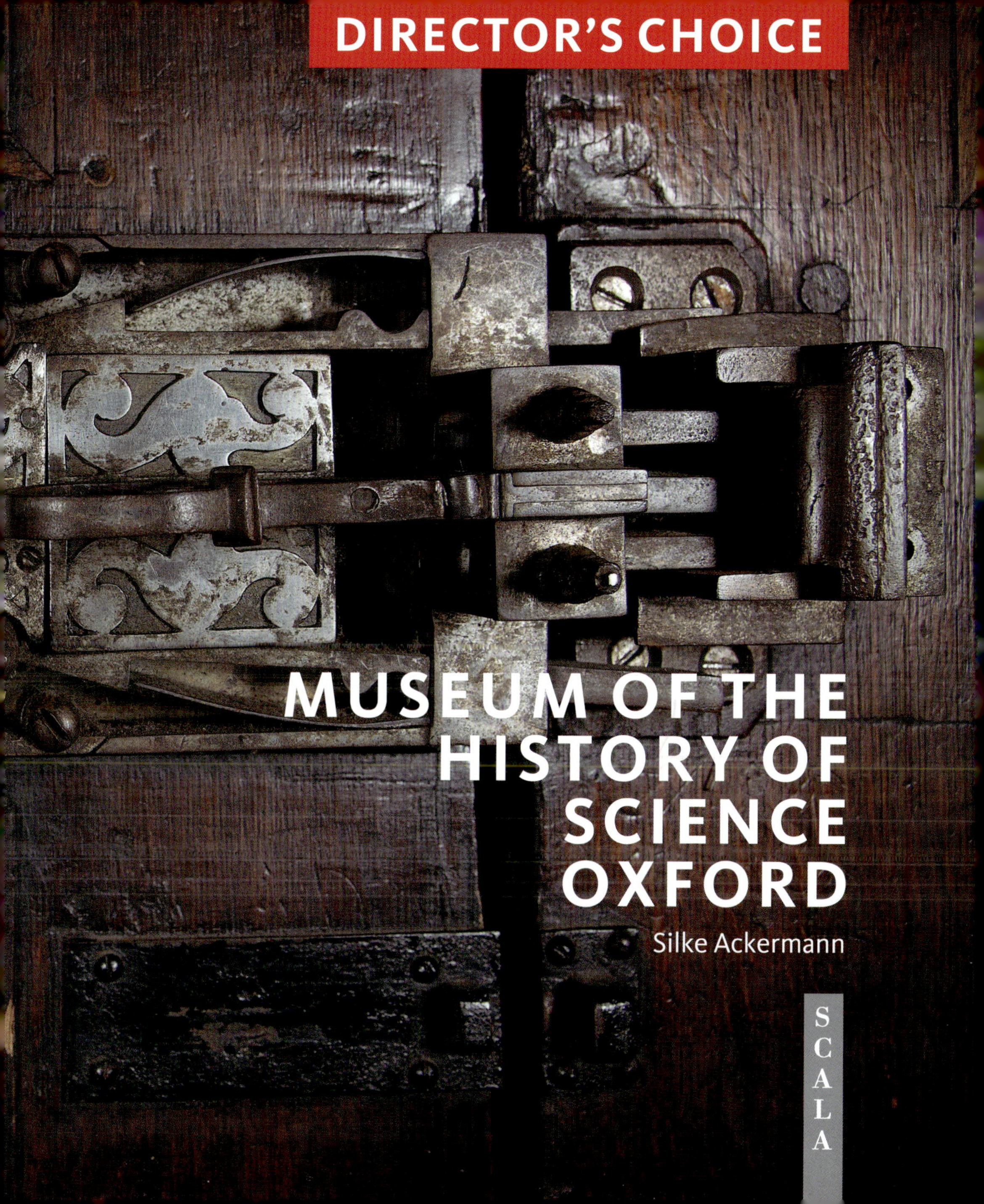

DIRECTOR'S CHOICE

MUSEUM OF THE HISTORY OF SCIENCE OXFORD

Silke Ackermann

SCALA

INTRODUCTION

A QUIRKY OLD BUILDING with a creaking staircase on Broad Street in the heart of Oxford, full of curious devices with unfamiliar names and even more obscure uses. Even its title, 'Museum of the History of Science', can be a challenge. 'Science' is in many ways a misleading term for the collections, a modern label that by definition excludes a wealth of topics that were key contributors to the knowledge of their day. The Museum includes subjects that are now seen as dangerous and dark, such as alchemy, astrology and divination, but also topics that are now too ordinary or vocational to count as science, whether surveying, drawing or designing sundials. But on closer inspection the smallest of the four university museums (the others are the Ashmolean, the Museum of Natural

Basement Gallery.

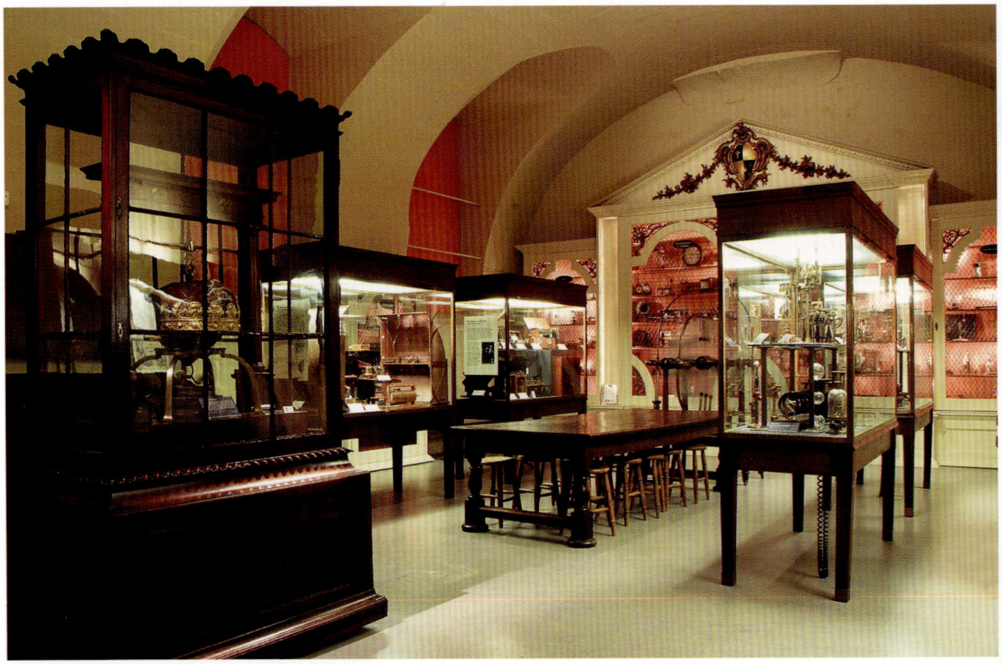

History and the Pitt Rivers Museum) reveals a treasure trove of artefacts that give an insight into the social and cultural context of 'science' in its widest sense and into the exchange of knowledge between cultures and generations.

For a long time in its history the 'Old Ashmolean Building', in which the MHS is housed, had precisely this knowledge exchange at its core. When it opened its doors in 1683 the Ashmolean Museum was the earliest purpose-built public museum in the world, housing the collections of the John Tradescants, father (?1570–1638) and son (1608–1662), and Elias Ashmole (1617–1692), on the top floor. But the building also encompassed a broad range of activities associated with the pursuit of 'natural knowledge'. 'The Museum' as originally conceived institutionalised a new way of learning about nature that emerged in the seventeenth century, with chemistry, anatomy and pharmacology being pursued in a laboratory in the basement and lecturing and demonstration taking place in the School of Natural History on the middle floor.

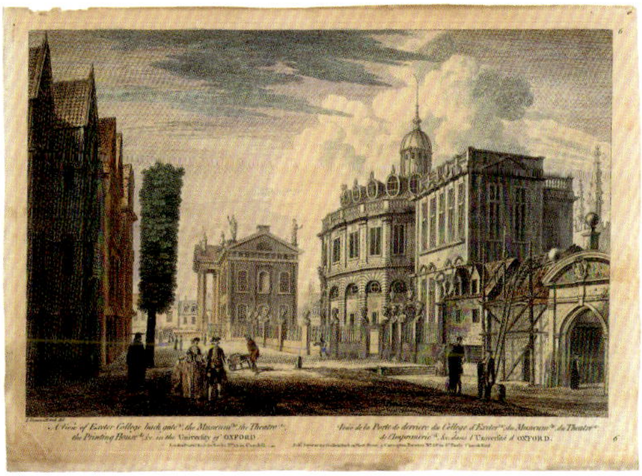

Print of Broad Street, showing the Clarendon Building, Sheldonian Theatre and Old Ashmolean, between 1767 and 1779 (inv. 40877).

All of this changed in the second half of the nineteenth century. In 1860 the zoological collections were moved to the new Museum of Natural History in Parks Road and the books were transferred to the Bodleian Library. The ethnographical collection left the building for the new Pitt Rivers Museum in 1886 and finally, in 1894, the remaining antiquities found a home in the new Ashmolean Museum, which was installed in the University Galleries' building in Beaumont Street. The old building served a variety of uses over the following decades.

A new lease of life as a museum began in 1924, thanks to the generous donation of Lewis Evans (1853–1930). Lewis was the second son of John Evans, one of the most distinguished Victorian archaeologists; his older brother Arthur, best known for his excavations at Knossos, became Keeper of the Ashmolean Museum in 1884, which he moved out of the old building only ten years later. The family fortune was based on paper making, enabling Lewis Evans to build up a unique collection of 'mathematical instruments', principally sundials and astrolabes with supporting textual sources. In 1922 he offered his collection to the University of Oxford on condition that they should be housed appropriately. It was the long history of scientific activities as much as its renowned status as 'The Museum' that persuaded Evans and his friend R.T. Gunther that this could only be the upper floor of the 'Old Ashmolean' building. Not without opposition, the 'Lewis Evans Collection' opened to the public on 5 May 1925.

Robert Theodore Gunther (1869–1940) became the founding curator of what was later, from 1935 onwards, known as the 'Museum of the History of Science'. Gunther was the son of a German curator (until 1916 the family name was 'Günther') who had played a prominent role in the late nineteenth-century development of the British Museum.

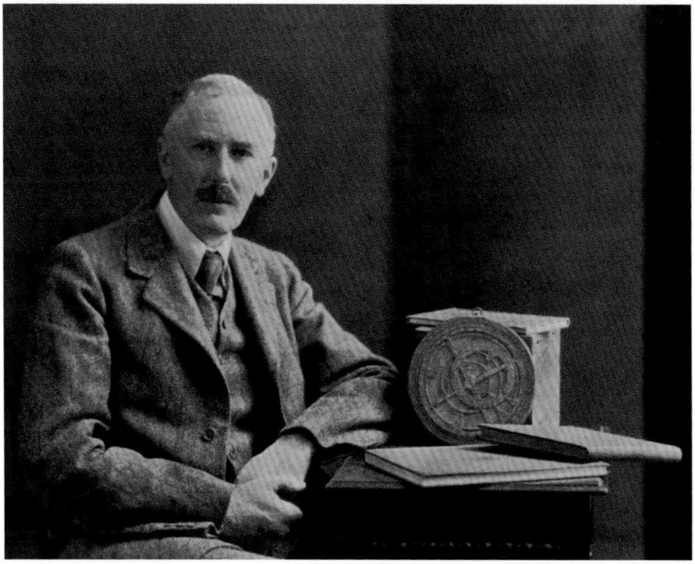

R.T. Gunther with a paper astrolabe from the MHS collection by John Prujean, Oxford, *c.*1680 (inv. 48821).

A zoologist by training, Gunther had embarked in the years following the First World War on a personal crusade to investigate and recover Oxford's forgotten history of scientific endeavour. This had a very particular and pointed significance in an Oxford often caricatured as an arts university. Poking into college cupboards and departmental back rooms, he discovered the material relics of a lost heritage. His notes on these finds – a novel sort of archaeology, as he conceived it – began to expand. He decided that, for the sake of its future preservation, this material needed to be published. And so began the grand enterprise of *Early Science in Oxford* (ultimately 14 volumes plus supplement, 1920–45) – parallel to steadily bringing together the evidence of scientific material culture that he had discovered. Gunther lobbied hard for the exclusive use of the building for the Museum, but it took another nine years after his death in 1940 before this was finally the case.

Following spectacular additions to the collections over the subsequent decades, the MHS now houses an unrivalled collection consisting of instruments, manuscripts and books, photographs and archival records from antiquity to the twenty-first century. About a fifth of the total collection of around 20,000 artefacts is currently on display.

Having myself only arrived at the MHS just over a year ago, the breadth of the collections and the stories behind individual artefacts can seem overwhelming. So many objects to choose, so many stories to tell! It has thus been a huge pleasure and privilege to discuss *their* favourite object(s) with members of our team, who have most generously shared their thoughts and insights with me and who contributed a wealth of knowledge, especially Katie Altizer, Elizabeth Bruton, Stephen Johnston, Tony Simcock and Sophie Waring. But without Keiko Ikeuchi's stunning photographs, the often unexpected beauty of science might have remained hidden.

Note

The entries that follow can cover only a small fraction of the information available on the selected objects. Readers wishing to find out about materials, dimensions, provenance or other details and stories can do so on the Museum's website: www.mhs.ox.ac.uk

His Name in Lights

Astrolabe for Shah Abbas II

Muhammad Muqim al-Yazdi
Persia, 1057AH 1647/8 CE
Inv. 45747

The Museum of the History of Science houses the world's largest and most important collection of astrolabes. With ancient origins and a two thousand-year history, the instrument illuminates astronomy, time-telling, astrology, mathematics and surveying across cultures, time and place.

An astrolabe combines heaven and earth. A rotating two-dimensional representation of the celestial sphere displays the motions of sun and stars, and the instrument captures their positions as seen from various locations on earth. The elaborate, web-like piece at the front, called the rete (from the Latin for 'web'), depicts a number of bright stars. Visible underneath is one of a number of plates, each marked for a specific latitude. A variety of scales on the front and back, together with a sighting device, enables a huge range of observations and complex calculations.

This astrolabe was made for the Safavid ruler Shah Abbas II (1632–1666) by Muhammad Muqim al-Yazdi in 1647/8, and given by the founding benefactor of the Museum, Lewis Evans. The delicate tracery in the upper part of the magnificent rete spells out the Shah's name in elegant calligraphy, placing him amongst the splendid lights of the sun and stars. This celestial connection is reinforced by the dedication that appears on the 'throne' of the astrolabe (the triangular part at the top):

> The supreme prince, the sultan, the most just, the greatest, lord of the centres of command, remover of the causes of tyranny and rebellion, king of the kings of the age, Abu al-Muzaffar Sultan Shah 'Abbas the Second, the Safawi, the Musawi, the Husaini, Bahadur Khan. May God Almighty perpetuate his Kingdom and his Empire and cause his justice and his benefits to spread over the worlds while the spheres revolve and the planets continue their courses!

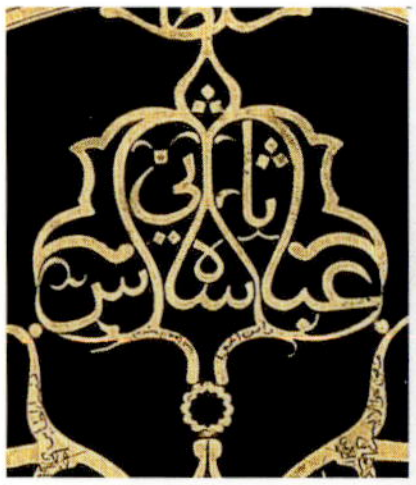

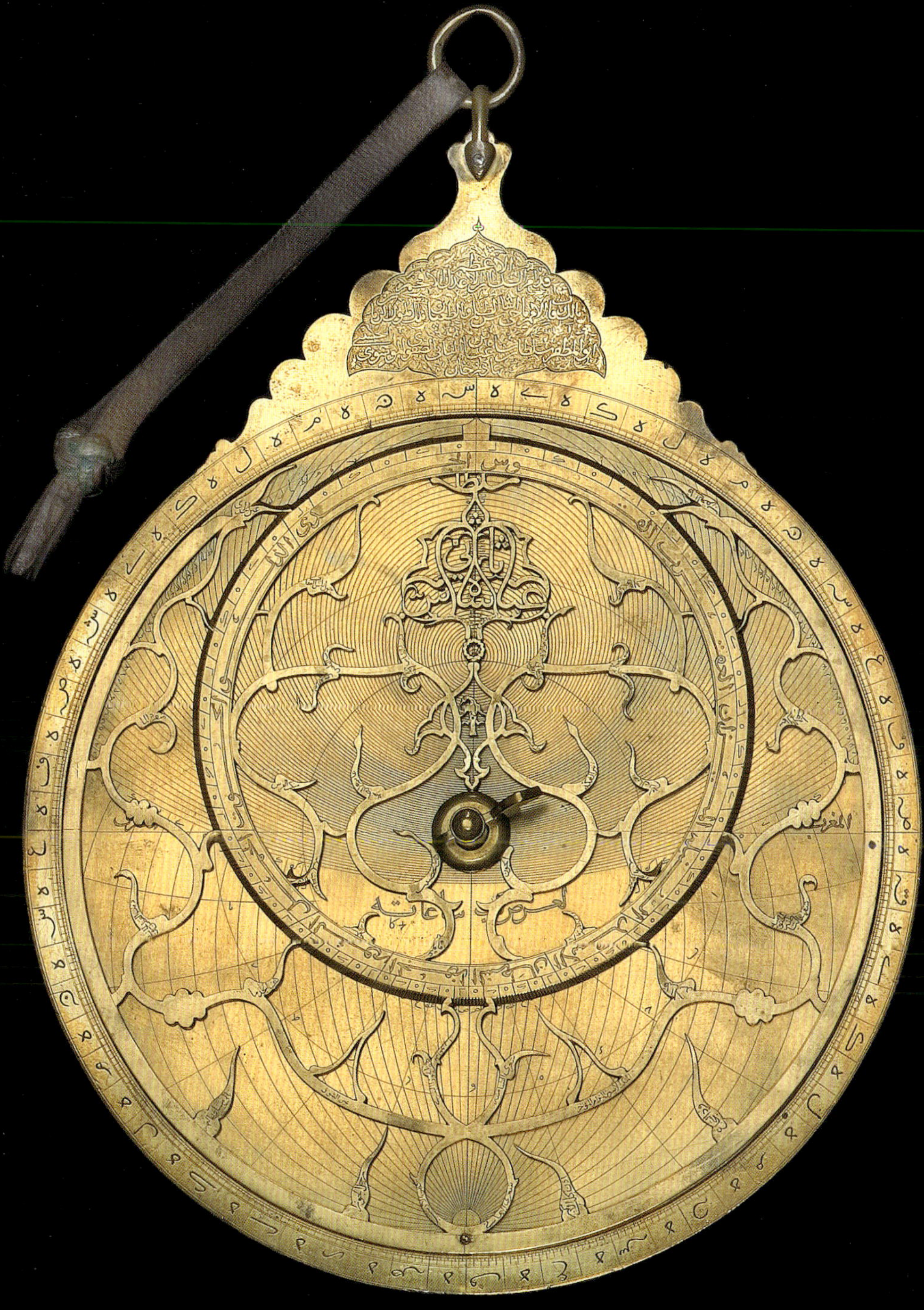

Portraying the Moon

Moon pastel and Selenographia

John Russell
London, 1795 and 1797
Invs 39338, 52085

This enormous depiction of the moon, over 5 ft in diameter, was completed in 1795 by the celebrated portrait artist John Russell (1745–1806). Russell was particularly noted for his skill in pastel and made his own crayons, achieving a variety of nuanced colours by blending just 14 pigments. While portraits dominated his public work, Russell's private efforts were dedicated to systematic astronomical sketches of the moon over a thirty-year period, with his interest most intense in the mid-1790s, when he obtained a Dollond achromatic refractor, the latest in telescope technology.

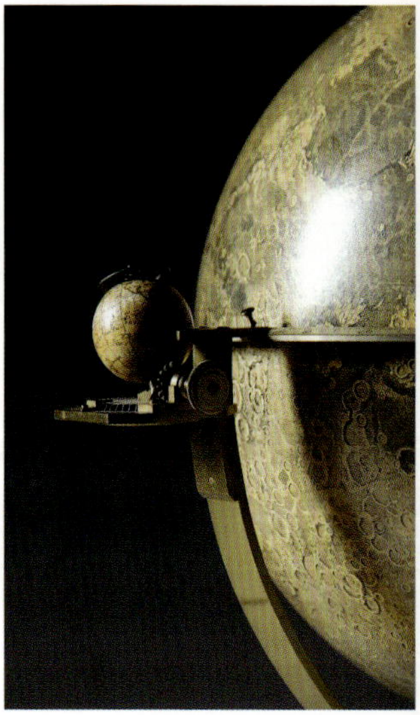

For Russell, celestial observation was an act of devotion. He noted in his diary, 'tonight being in the Street I had a sight of the Stars that God was pleas'd to preach to me from, and I had my Soul filled with the Hopes of ... Immortality'. But this was not just religious and artistic worship. Like other contemporary astronomers, Russell was systematic in his methods, observing and charting the moon during his midnight vigils. Russell completed 187 moon drawings, and this obsessive and devoted study resulted in both this large pastel and the moon globe, which Russell named the 'Selenographia'.

Unlike the lunar pastels and sketches, which were private, the globes were made for the commercial market and could be purchased by subscription. Russell designed a stand for the globe, which demonstrated the libration, or motions, of the moon. The surface of the globe was printed from engraved copper plates and the sheets of paper cut into so-called gores that were then attached to the papier-mâché globe. Only a few examples of John Russell's Selenographia are known to have been sold and even fewer survive.

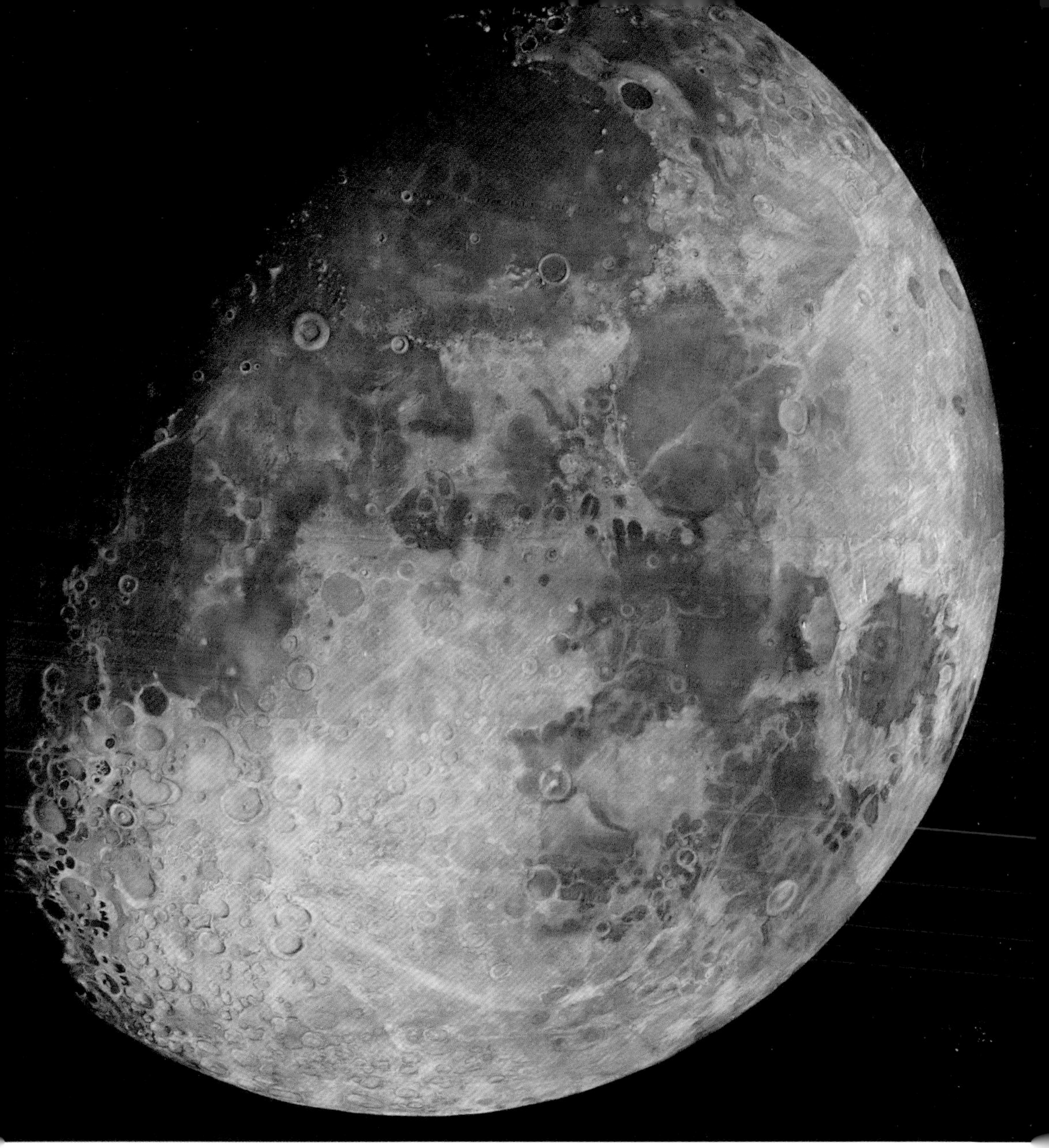

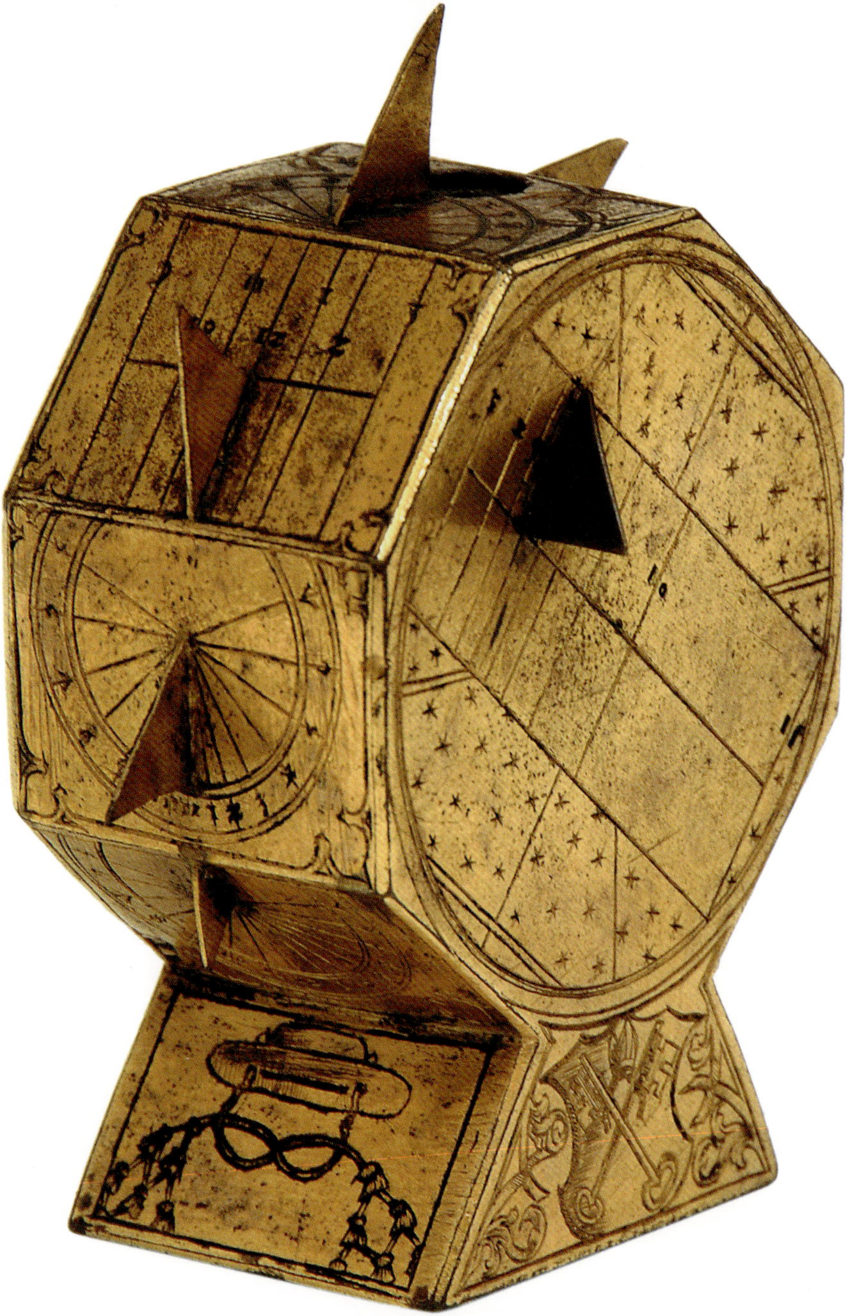

Ingenuity and Ownership

Polyhedral sundial

Nicolaus Kratzer
Oxford or London, between 1518 and 1530
Inv. 54054

As a figure of popular imagination, Henry VIII is widely remembered for his six wives and the drama of their different fates. But his reign was about much more. Under his rule the Court looked increasingly towards Continental Europe for intellectual and aesthetic innovation, consciously emulating the ideals of contemporary Renaissance courts.

One of the many notable arrivals from across the Channel was the German mathematician Nicolaus Kratzer (*c.*1487–1550). He became Court Astronomer to Henry VIII and is now most vividly remembered through his portrait by the celebrated painter Hans Holbein (1497–1543). Holbein shows Kratzer not as a bookish figure but as a practising instrument maker, working on a polyhedral sundial, surrounded by tools and mathematical instruments.

Holbein's portrait reminds us that the mathematics of the sixteenth century could be a surprisingly manual affair. It also sets the scene for this small gilt-brass instrument, another sundial in elaborate polyhedral form. Although not signed, it has been attributed to Kratzer, in part because it was evidently made for Cardinal Wolsey (1473–1530), one of Henry VIII's most trusted advisors until his fatal fall from favour.

Wolsey had appointed Kratzer to lecture on mathematics in Oxford and this sundial was presumably created as part of that personal connection. The base of the instrument carries Wolsey's coat of arms as well as those of the Archbishop of York (a position he attained in 1518); two cardinal's hats complete the decoration. It suggests a pleasure in the ingenuity of mathematics, with nine different dials arranged on an octagonal body whose irregular angles show that it was intended for use at a latitude of about 52 degrees, just right for Oxford.

Calculating by Steam

Parts of difference engine

CHARLES BABBAGE
London, *c.*1822–30
Inv. 94229

AT A MEETING of the newly formed Royal Astronomical Society in 1822, Charles Babbage (1791–1871) announced that he could build a calculating engine that would produce astronomical or mathematical tables by steam power. Tables were vital for science, education and trade and were calculated by human computers. Inevitably they were subject to errors and Babbage claimed that his new machine, the difference engine, would completely eliminate these errors and make the production of the tables much quicker.

The potential of the engine to improve safety and cut costs attracted the attention of the British government, which, even after the end of the Napoleonic Wars in 1815, was still fiercely competing with France, but through trade rather than war. The government offered grants for any technology or method able to improve navigation. Babbage started working on his difference engine immediately and the project ran for twenty years. Despite receiving the extraordinarily large sum of £17,000 in government funds, Babbage never completed his difference engine.

Several assemblies of difference engine parts have survived. All have a wooden mounting except for ours, which suggests that it was not one of those put together by Babbage's son Henry. It most likely came directly from Babbage's workshop and was given by him to his friend Harry Wilmot Buxton (1805–1880), along with a collection of manuscripts. The manuscripts and the engine assembly were given to the Museum by Buxton's great-granddaughter. Included in this collection are the first-ever thoughts committed to paper by Babbage about how the difference engine would work, scribbled down after a conversation with his good friend John Herschel.

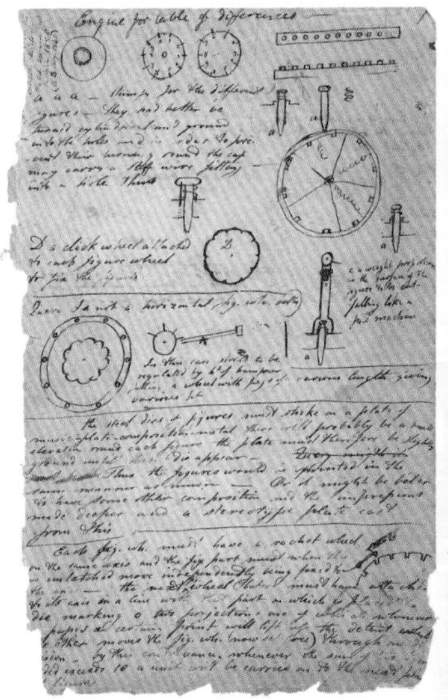

Babbage's very first notes and designs for the difference engine, 1821 (MS Buxton 9).

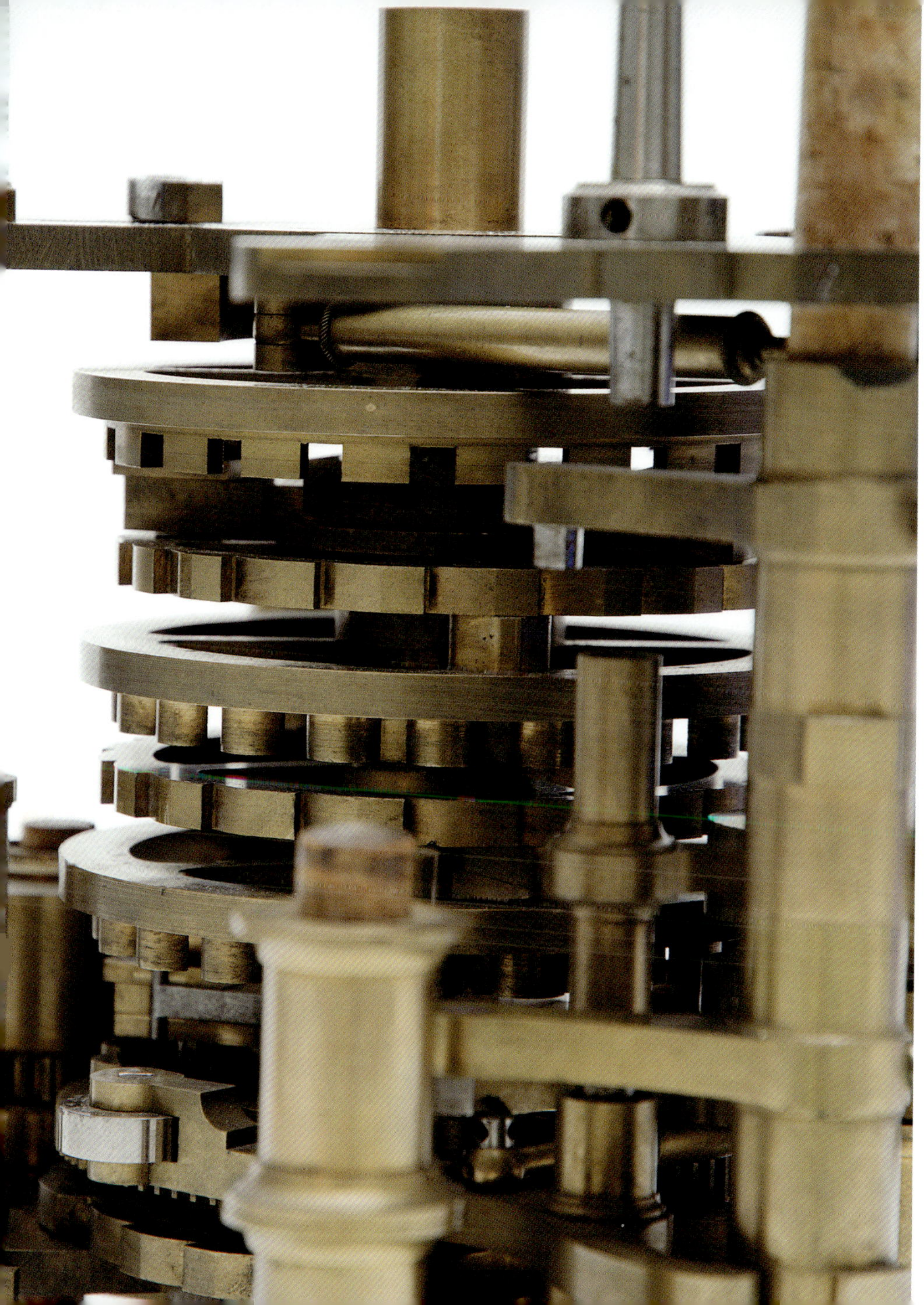

Saving Countless Lives

Original penicillin culture and specimen

Howard Florey and colleagues
Oxford, *c.*1940–1
Invs 23034, 14439

The discovery of the antibacterial properties of penicillin by Alexander Fleming (1881–1955) is legendary, but the wartime transformation of this breakthrough into a clinical product for patients was achieved at the Sir William Dunn School of Pathology, Oxford. Howard Florey (1898–1968) led a team including Norman Heatley and the German refugee Ernst Chain, which laid the basis for an industry of antibiotic production that transformed Fleming's biological curiosity into a tool that fundamentally changed medicine.

Penicillin is produced by a fungus that can be cultured. The dried-up contents of the cotton-sealed flask are one of the original cultures that Florey's team used to extract penicillin. Early tests on mice were very promising, resulting in Florey's laboratory being transformed into a production centre to make enough penicillin for human clinical trials in 1941. Two thousand litres of culture fluid were required to obtain enough penicillin to treat a single case of sepsis. The team therefore worked hard to increase the yield from the cultures. So precious were early stocks of the drug that none could be wasted. When administered, not all the penicillin was metabolised and some could be recovered from the urine of patients and reused, as witnessed by this tiny glass vial dated 25 June 1941.

Almost miraculous recoveries from apparently fatal infections demonstrated the therapeutic value of the antibiotic. Wartime Britain was unable to produce penicillin on the scale so urgently required, so Florey and Heatley made a secret visit across the Atlantic in 1941 to mobilise American interest. Pharmaceutical companies in the United States banded together to devise and develop a new, much higher-capacity industrial process, which was used to save countless lives.

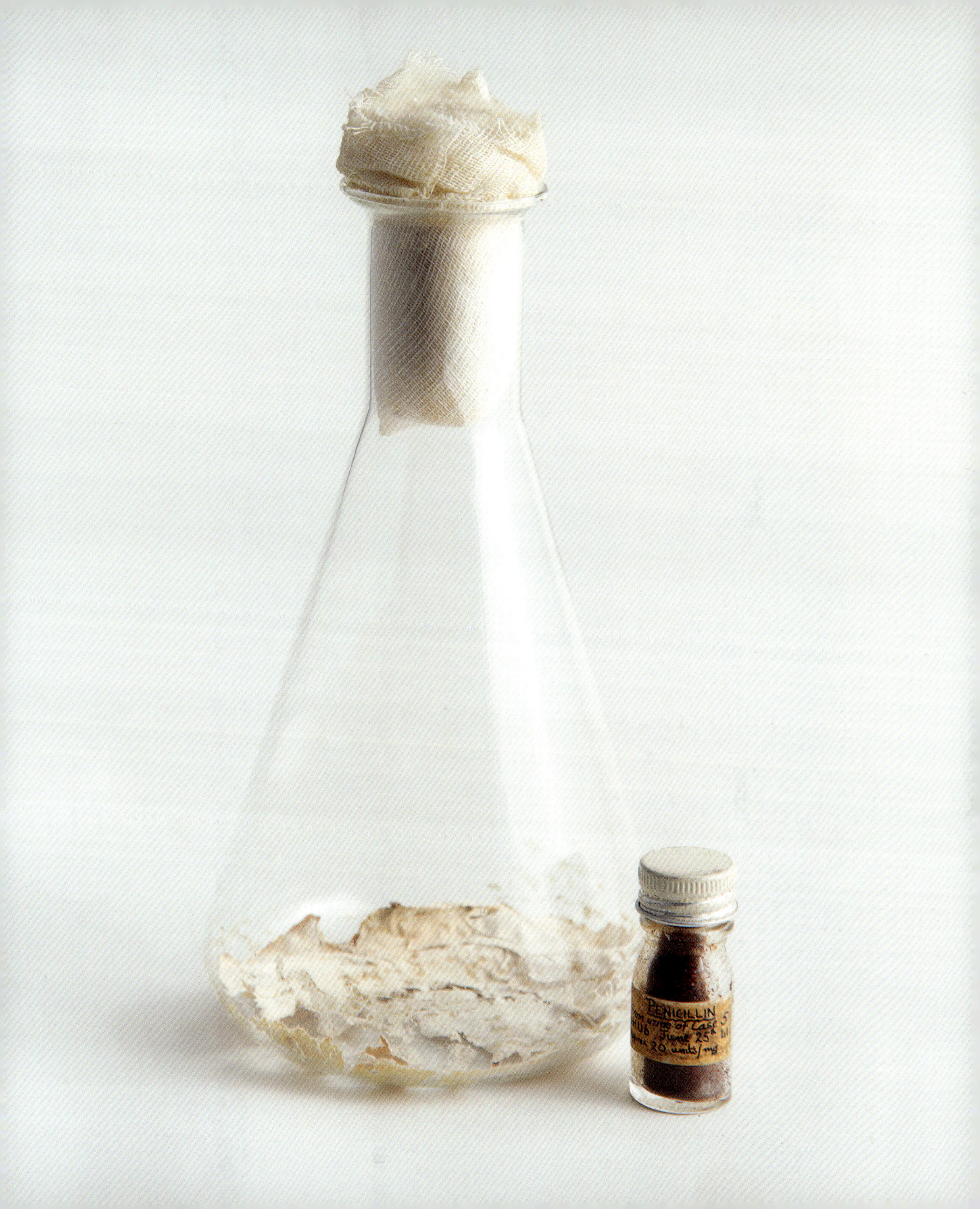
PENICILLIN
urine of Case 5
June 25
20 units/mg

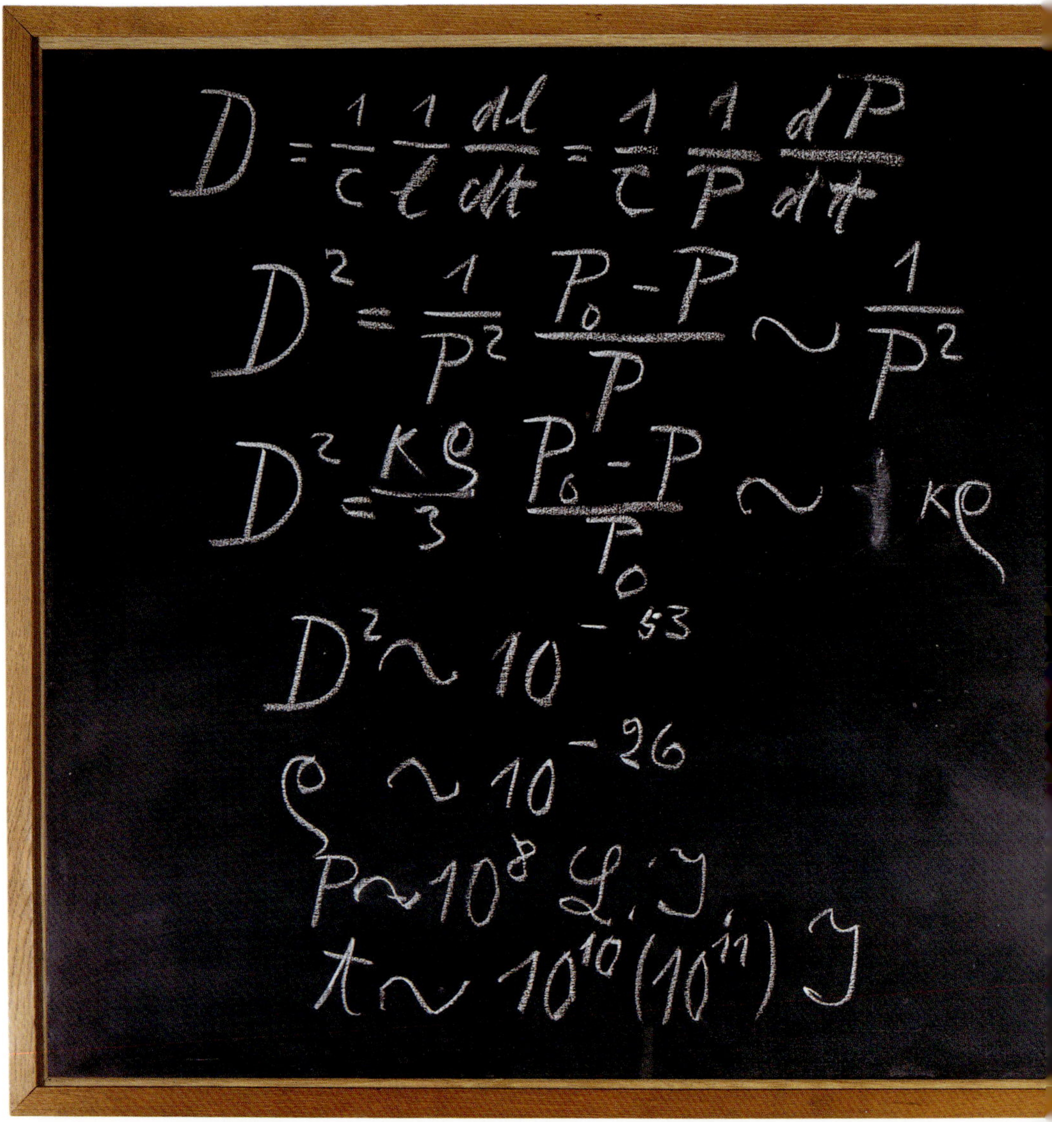
D = 1/c 1/l dl/dt = 1/c 1/P dP/dt
D^2 = 1/P^2 (P_0 - P)/P ~ 1/P^2
D^2 = κρ/3 (P_0 - P)/P_0 ~ κρ
D^2 ~ 10^-53
ρ ~ 10^-26
P ~ 10^8 L.J
t ~ 10^10 (10^11) J

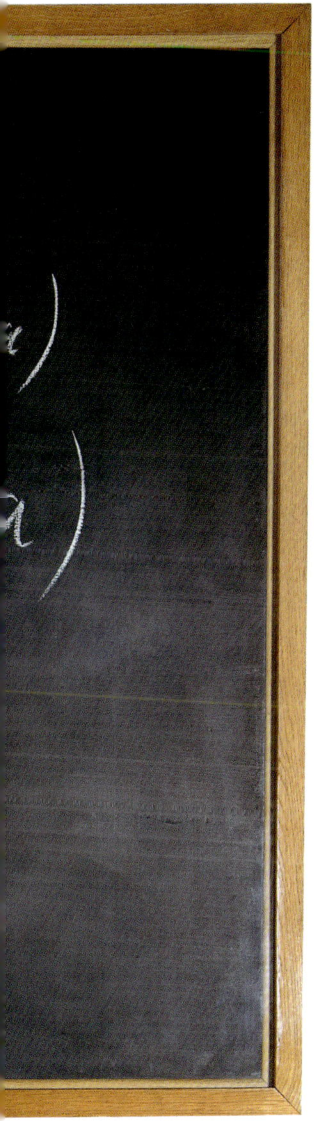

Relic of a Secular Saint

Blackboard used by Albert Einstein

Oxford, 16 May 1931

Inv. 44725

Albert Einstein (1879–1955) was already internationally celebrated for his special and general theories of relativity when he was invited to Oxford in 1931. This was a time of great excitement in the astronomical world and *The Oxford Times* reported that the audience for his first lecture was full to capacity: 'Large crowds of undergraduates and senior members of the University assembled ... the audience numbered over 400 including university professors, Heads of Houses and several noted scientific scholars, and when the lecture began a number of people were standing at the back of the hall.'

This blackboard was preserved from Einstein's second lecture, on 16 May 1931, and its equations connect the age, density and size of the universe. During the 1920s the work of the astronomer Edwin Hubble (1889–1953) on red shifts had established that other galaxies were receding from our own. In combination with his own theories, Einstein was able to use Hubble's observations to explain this apparent expansion of the universe. The mathematical models were very complex, but at about this time Einstein was developing a much simpler model, in which troublesome factors were ignored. This is the subject that the blackboard records.

It is a sign of Einstein's remarkable international fame that his chalked equations were not rubbed out at the end of the day. A blackboard should be the opposite of a permanent record. But instead this board was brought directly to the Museum, where it has been preserved ever since. Some visitors today treat it almost as an object of veneration, anxiously requesting its location on arrival and eager to experience some connection with this near-mythical figure of science.

Experiments with Angels

Marble copy of John Dee's Holy Table

England, mid-17th century

Inv. 15449

In his own lifetime the sixteenth-century mathematician John Dee (1527–1609) enjoyed an international reputation as a scholar and acted as advisor to Queen Elizabeth I. However, he is now mostly famous – or rather notorious – for his 'angelic conversations', in which with the help of his medium, or 'skryer', Edward Kelley (1555–1597/8), he conjured a series of angels in the years between 1582 and 1589.

This was no fringe pursuit: during the Renaissance angels were believed to retain knowledge of the supernatural reality underlying the physical world of God's creation. Adam, and therefore mankind, had lost this knowledge when expelled from the Garden of Eden. Dee believed that through interactions with angels he could obtain a true understanding of both the spiritual and natural realms. For him, angelic communication was as viable a method as mathematics for understanding the natural world.

Under instruction from the archangel Uriel, a brightly painted wooden table was created on which the characters of the 'Enochian' alphabet were inscribed. Angels could simply indicate successive letters to convey messages, which Kelley then interpreted for Dee. The original wooden table is lost; this marble copy, made nearly a century after Dee's work, was based on a published engraving that appeared in 1659, in a volume containing detailed transcripts of Dee's angelic consultations.

The marble copy may have been owned by the astrologer William Lilly (1602–1681), the author of a series of popular astrological almanacs and an authority on more recondite questions of angels, spirits and fairies. The copying of a Renaissance artefact shows that an interest in the occult continued into the seventeenth century, a period traditionally seen as witnessing the rise of inductive and experimental approaches to the study of the natural world.

The HOLY TABLE

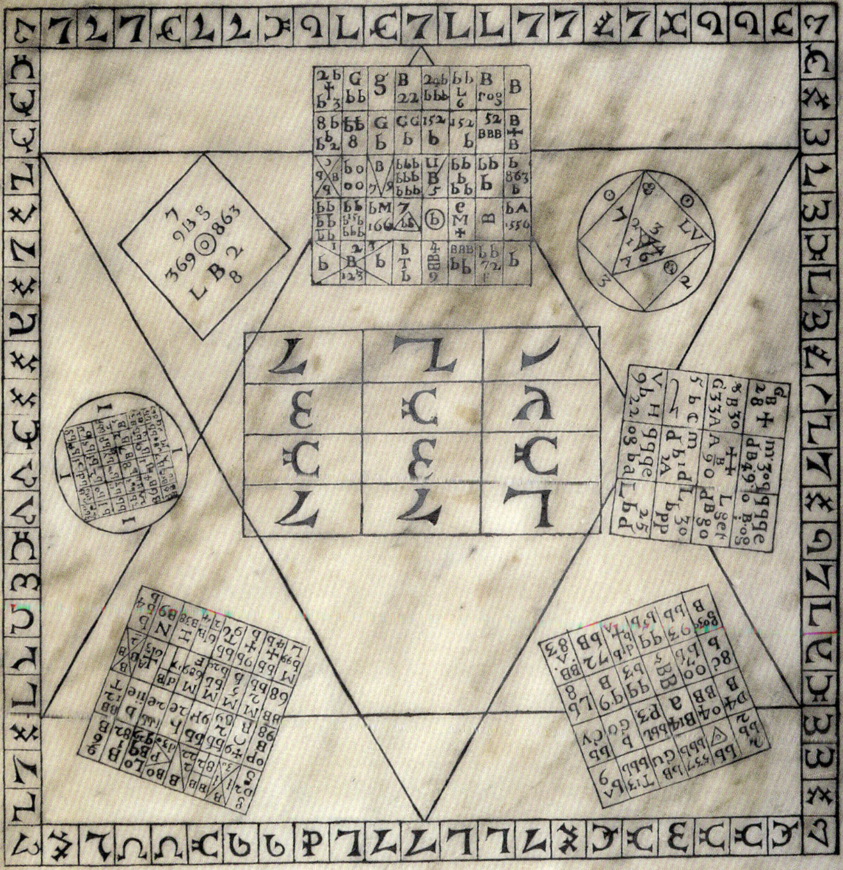

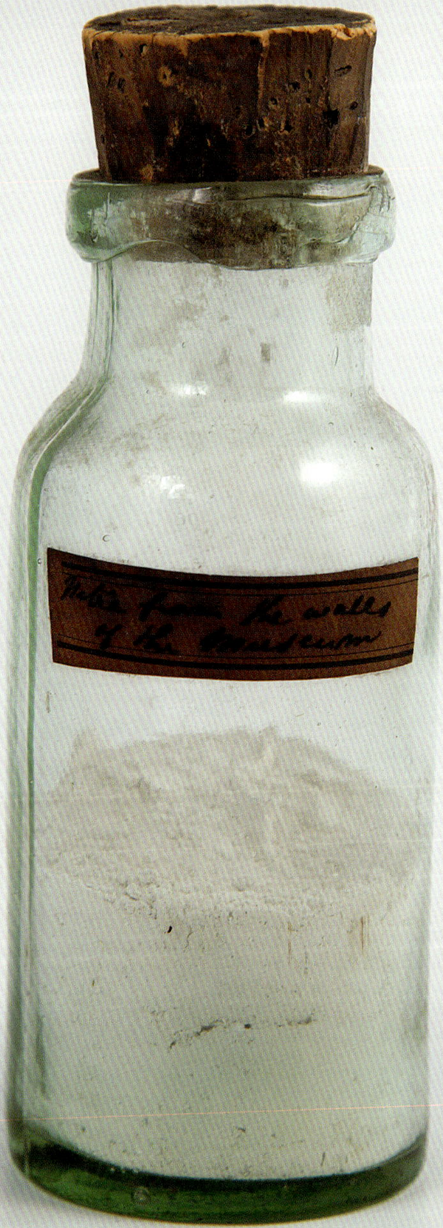

The Chemical Museum

Nitre efflorescence scraped from the Museum's basement walls

Oxford, *c.*1814

Inv. 41667

THE ASHMOLEAN MUSEUM opened in 1683 in the building now occupied by the Museum of the History of Science. For more than 150 years the main basement gallery was a chemical laboratory, used for teaching and research and even, in the Ashmolean's early years, as a place for dispensing chemical medicines prepared on the premises.

Curiously, the building itself became an object of chemical study. The corked glass bottle shown here is labelled 'Nitre from the walls of the Museum' and contains a specimen of the mineral efflorescence that continues to grow on the basement walls even now.

A succession of Oxford chemistry professors in the eighteenth and early nineteenth centuries published analyses of the composition of this efflorescence. This particular specimen may have been collected by Dr John Kidd (1775–1851), the university's first Aldrichian Professor of Chemistry, whose analysis appeared in the *Philosophical Transactions of the Royal Society* in 1814.

One of the earlier accounts from 1785 gave, in very circumspect language, the reason for the nitrogren-rich content of these wall scrapings:

> The wall of one end of my chemical school, or laboratory, is almost entirely covered with such an efflorescence. The laboratory is a large vaulted room under ground, into which the sun seldom has admission. It is built of stone, and therefore, except when fires are kept up for lectures, or occasional experiments, is liable to be damp. The wall, to which I allude, is immediately under a retired passage, a very convenient place of retreat to foot-passengers under certain circumstances of necessity. The ground, therefore, and the adjacent wall have been for years largely impregnated with excrementitious animal fluids, in all the different stages of putrefaction.

The Spectacle of the Magnet

Natural magnet

Mounted by William Dugood
London, *c.*1728
Inv. 47759

At the heart of this fine piece of eighteenth-century furniture is a natural magnet, a large lodestone dug out of the ground. The whole represents a remarkable survival: a scientific object in its original museum display case.

Lodestones were once prized for their seemingly magical properties: what better example of the occult operations of nature than a magnet invisibly attracting iron? By the eighteenth century, natural philosophy had tamed – even if not fully explained – the hidden power of the lodestone. But a sense of wonder remained. A particularly large and powerful specimen was therefore an appealing object for a noble connoisseur. William Cavendish, 2nd Duke of Devonshire (1672–1729) acquired this example and employed a jeweller to mount it, disguising the irregular and rather undistinguished piece of ore under a gilt ducal coronet.

Magnetically attached under the crown is a triangular piece of iron with a hook, on which can be progressively hung a set of lead weights. When gravity finally overcomes the power of the magnet the weights fall dramatically on to a sprung leather pad, no doubt with a satisfyingly conclusive sound.

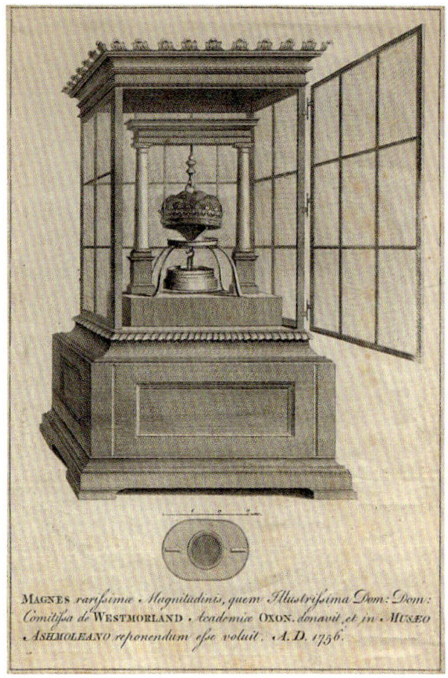

Engraving of the Countess of Westmorland's lodestone (inv. 13671).

The Duke's lodestone was not just for private entertainment but was exhibited for experiments at the Royal Society in London in 1730. The apparatus was kept in the Cavendish family and given to the Ashmolean Museum in 1756 by the Countess of Westmorland, along with the addition of the accompanying large showcase from her husband. This aristocratic donation was so notable that an engraving was produced to commemorate the event.

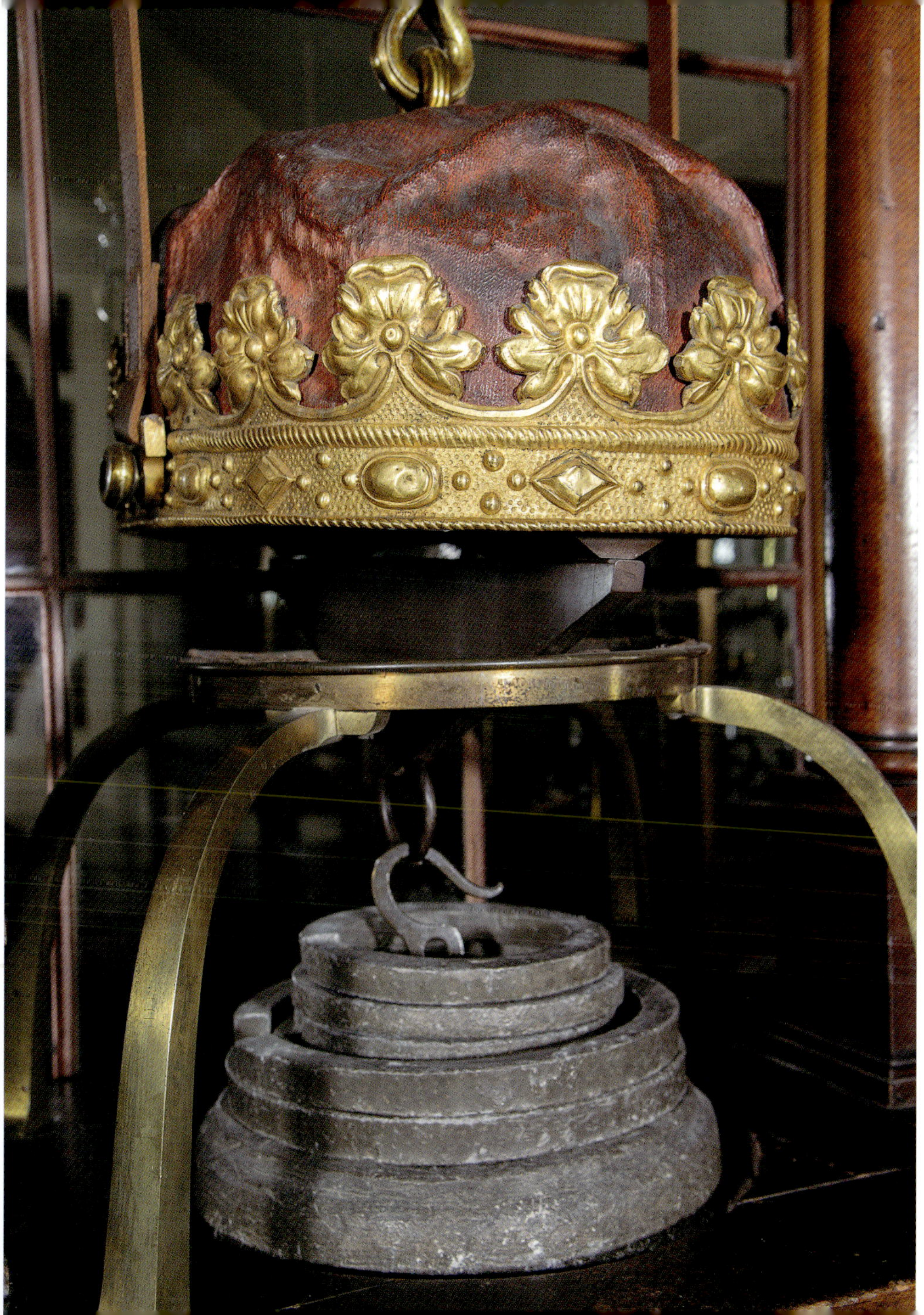

Fit for a Queen

Astrolabe

THOMAS GEMINI
London, 1559
Inv. 42223

THE DESIGN of this brass astrolabe, which was once magnificently gilt, is unusually fine. Each star pointer on the rete (the web-like disc at the front) terminates in a sculpted five-pointed star and the connecting strapwork consists of elements apparently passing over and under each other in three dimensions. A tour de force of decorative art, the rete also features the distinctive tulip shape characteristic of astrolabes from the contemporary Arsenius workshop of Louvain. Although created in London, the instrument was actually made by a Flemish immigrant, Thomas Gemini (*c.*1510–1562), who wittily signed himself with the zodiac sign 'Thomas ♊ 1559'.

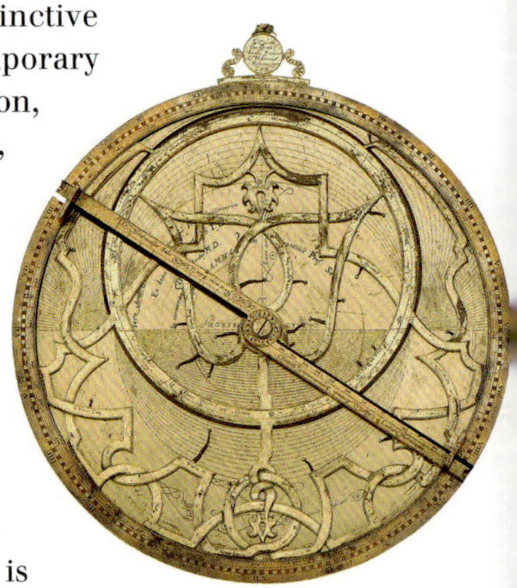

The disc at the top of the instrument is inscribed on one side with the coat of arms of Queen Elizabeth and on the other with a Latin inscription which translates as 'Elizabeth by Grace of God Queen of England, France and Ireland'. This is the same text that appears impersonally on her coinage; here, however, it most likely indicates that this was her very own astrolabe, whose quality and size were indeed regal. The supporting evidence is circumstantial but very suggestive and appears in some financial accounts of Elizabeth's court favourite, Sir Robert Dudley (1532–1588), the future Earl of Leicester. A month after being appointed a Knight of the Garter in April 1559, and just before going to Windsor for the ceremony, Dudley's accounts record a payment 'to Jemynie the Frencheman for an instrument of astronymye £10'. That sum is double what a skilled artisan would earn in a year, so this was an expensive commission. Surely the astrolabe was Dudley's gift to the Queen, intended to cement his place at court and offered in return for his investiture in England's highest chivalric order?

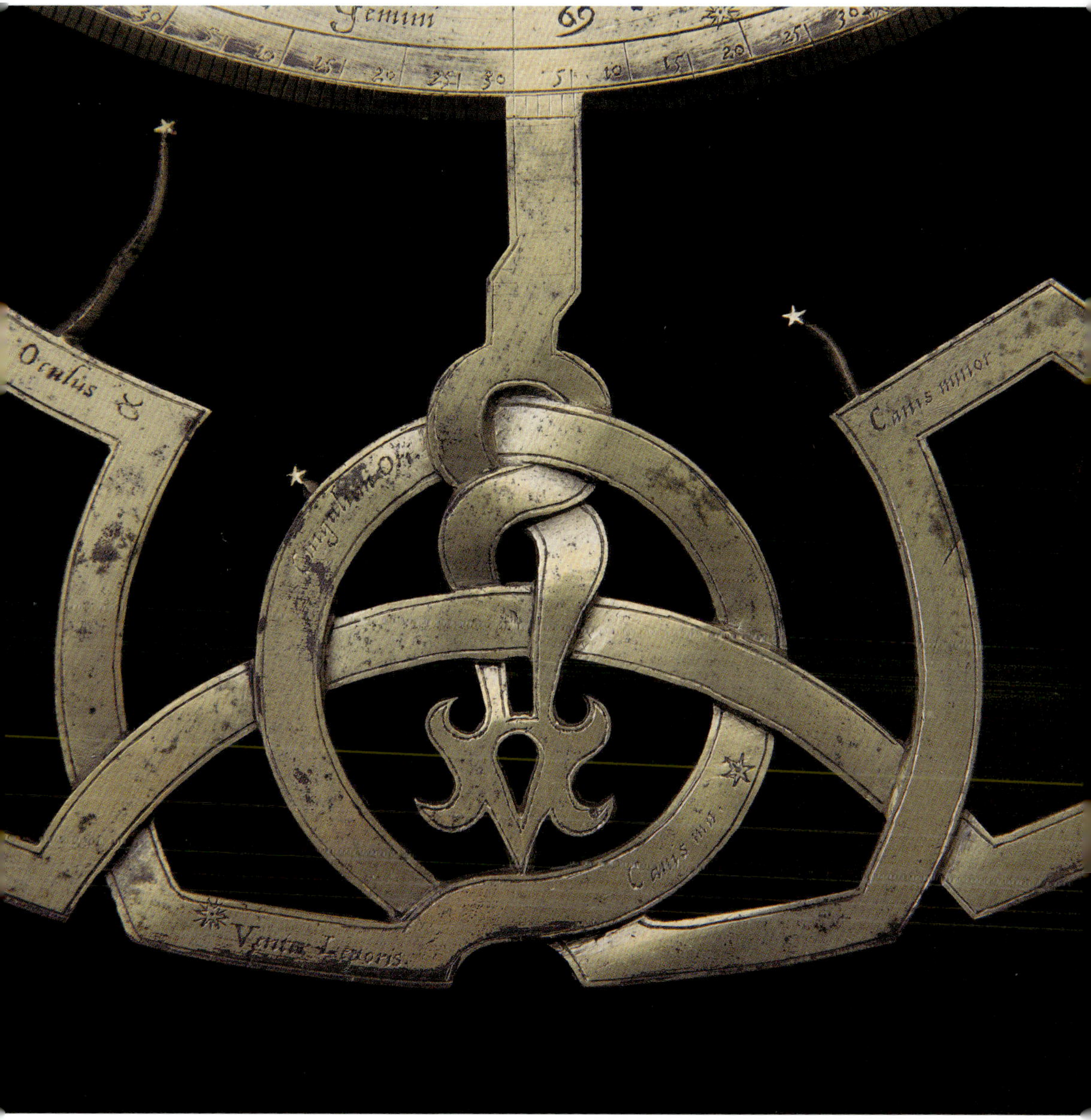
Oculus
Canis minor
Venter Leporis

Wondrous Wheelwork

Geared astrolabe

Muhammad Ibn Abi Bakr
Isfahan, 618 AH 1221/2 CE
Inv. 48213

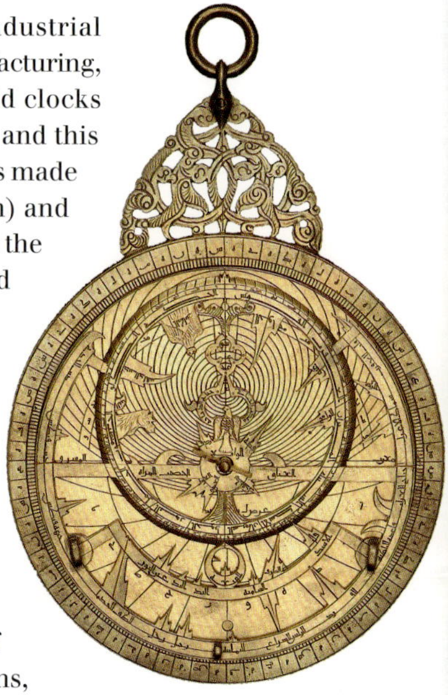

Our first thought of gearwheels might be in an industrial machine and associated with modernity, power and manufacturing, or perhaps a clock, with a longer history. But where did clocks come from? They developed from astronomical devices and this is the oldest complete geared machine in the world. It was made in early-thirteenth-century Isfahan (in modern-day Iran) and is part of a largely lost tradition going all the way back to the more complex but fragmentary remains of the celebrated ancient Greek Antikythera mechanism.

An astrolabe on one side, the reverse carries three displays that are driven by the gearing hidden inside the body of the instrument. (Our main image is a composite to show some of the wheelwork.) At the top is a circular window showing the phase of the moon, and beside it a smaller rectangular window with the age of the moon in the days of its monthly cycle. Underneath are two rings, one for the sun and the other for the moon, showing their motions against the circle of the zodiac and revealing their conjunctions and oppositions, and thus the possibility of eclipses.

The inclusion of the internal gearing created an astrolabe considerably thicker than normal. This gave the maker an unusual opportunity for decoration: the rim of the instrument carries images of the zodiac signs in inlaid silver, separated by mysterious figures whose iconography has not been fully deciphered. The long inscription around the circumference of the back ends with: 'Behold the disc! It will show you many of the wonders that prove the wisdom of the Merciful; its different motions are by virtue of a single mover, and it has meanings going beyond all meanings.'

OPPOSITE:
The back of the astrolabe with the left-hand half of the image showing an overlay of the gearing hidden inside the instrument.

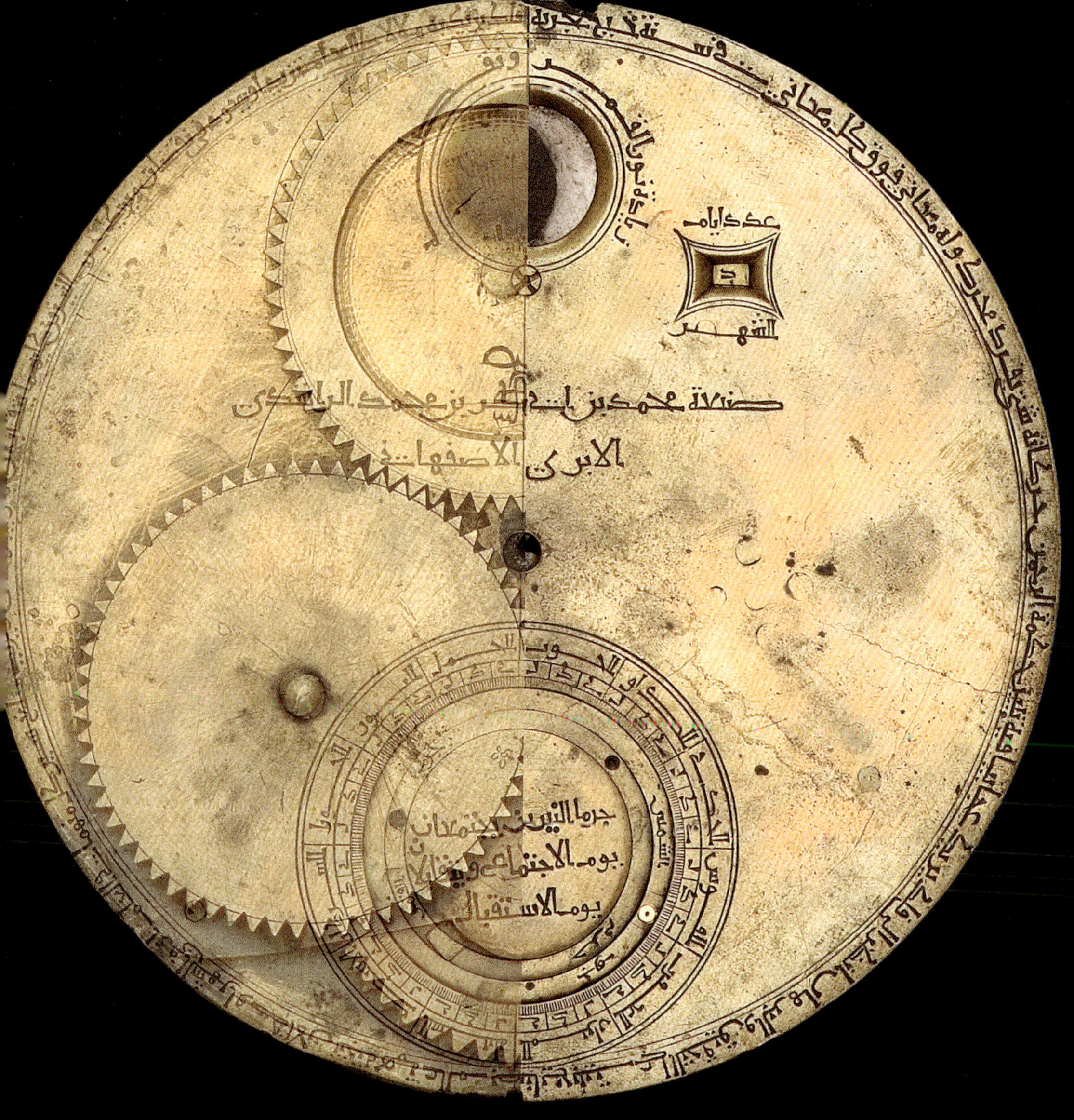
عدد ايام
الشهر

Photographing Wonderland

Lewis Carroll's wet-collodion outfit of photographic chemicals

HOCKIN & CO.
London, *c.*1860
Inv. 61498

Photograph (albumen print) of Annie Rogers and Mary Jackson as Queen Eleanor and Fair Rosamund

C.L DODGSON
Oxford, 1863
Inv. 30302

THE LID of this wooden box carries the initials C.L.D for Charles Lutwidge Dodgson (1832–1898), who is best known as Lewis Carroll, the author of *Alice's Adventures in Wonderland* (1865) and *Through the Looking Glass* (1872).

Dodgson was a clergyman and a mathematician at Christ Church, Oxford. In 1856, at the age of 24, he took up photography, then still a new art form. He soon became a well-known gentleman-photographer, finding it a useful entrée into higher social circles. He made portraits of many notable sitters from the arts and sciences, such as John Everett Millais, Ellen Terry, Dante Gabriel Rossetti, Julia Margaret Cameron, Michael Faraday, Lord Salisbury and Alfred, Lord Tennyson. He also photographed local figures, including the daughters of the Vice-Chancellor of the University of Oxford, Henry Liddell. Alice Liddell became the inspiration for Alice in Wonderland when she asked Dodgson to tell her a story during a boat trip on the River Thames.

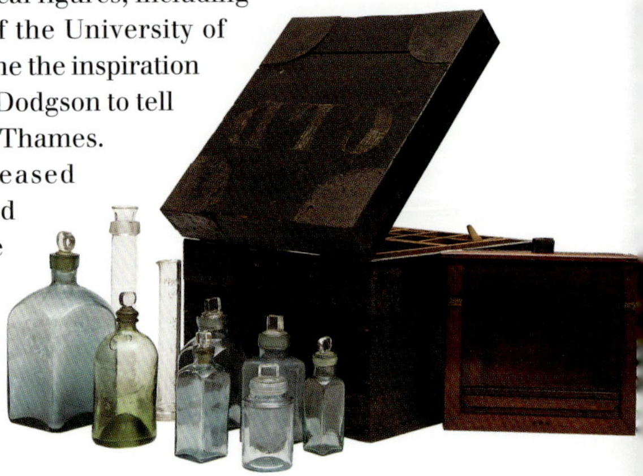

By the time Dodgson abruptly ceased photography in 1880, he had established his own studio and created around three thousand images, though fewer than a thousand survive. The chemicals contained in this box were used both to prepare the photographic glass plates before taking a picture, and also to fix and wash the image afterwards.

The Museum has one photograph by Dodgson, featuring a famous Oxford legend. Probably taken on 3 July 1863, it shows Annie Rodgers on the right dressed as Queen Eleanor, the wife of Henry II, and Mary Jackson on the left as Fair Rosamund, his mistress. Eleanor offers her rival the choice of a dagger or poison, while Fair Rosamund kneels, her hands clasped.

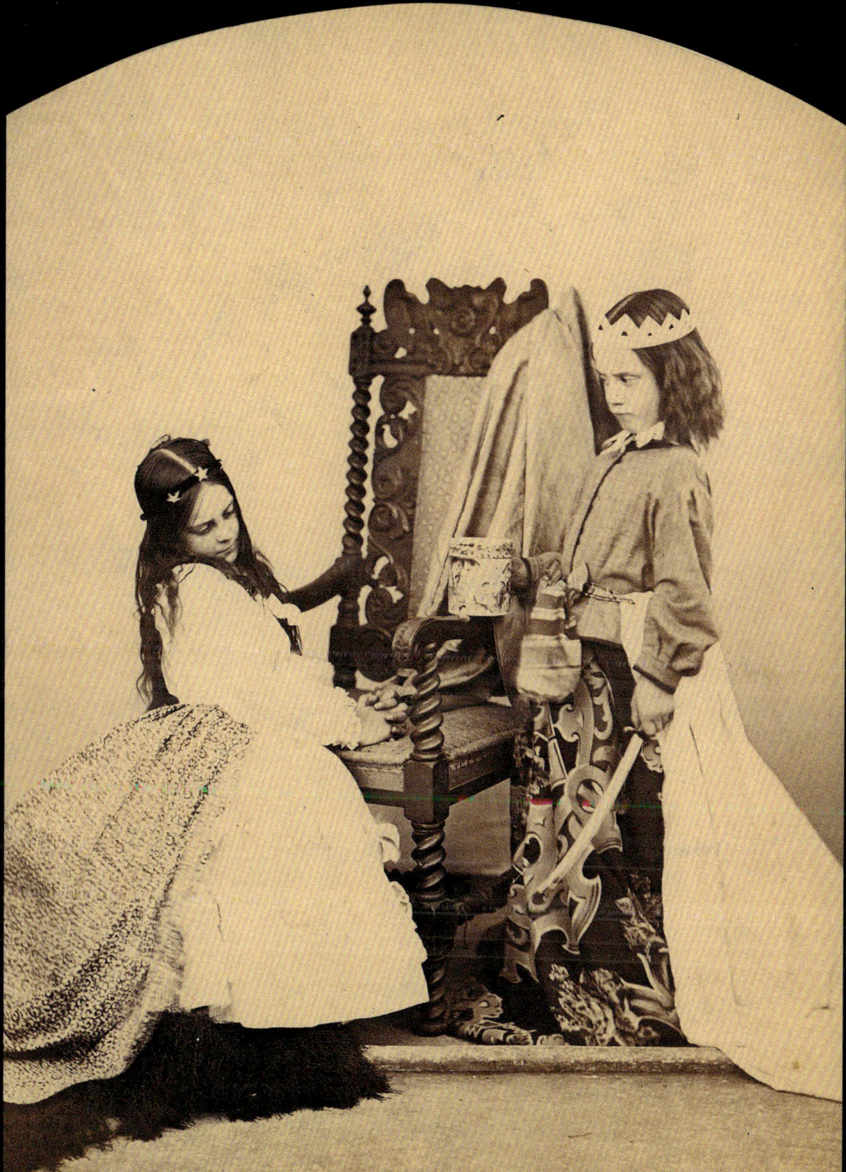

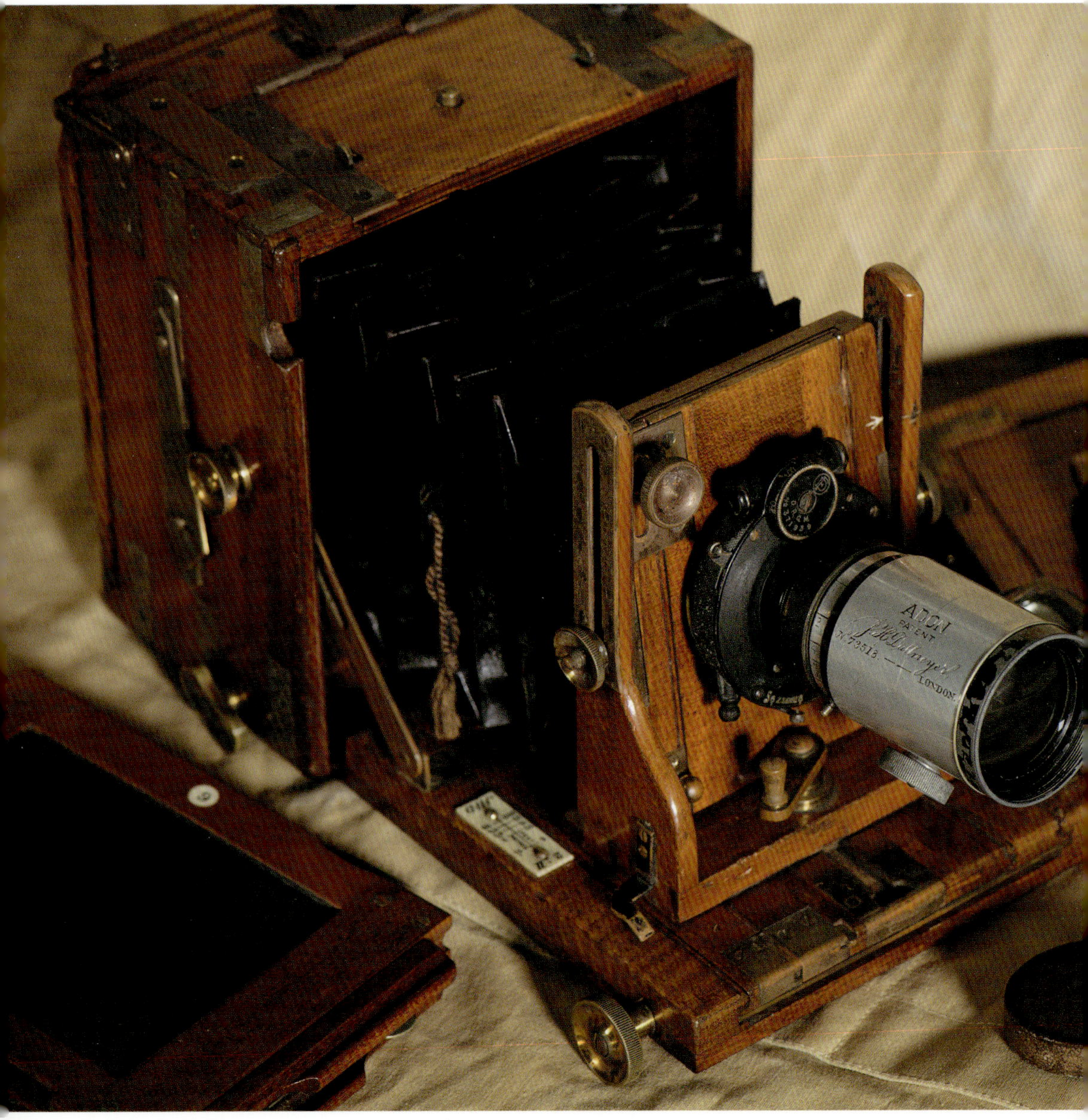
APLAN
PATENT
J.H. Dallmeyer
LONDON

Our Man in the Middle East

Lawrence of Arabia's camera

J.H. Dallmeyer
London, 1910
Inv. 21126

Thomas Edward Lawrence (1888–1935) is remembered for his dramatic career as military strategist, guerrilla and diplomat in Arabia during and after the First World War. Lawrence's in-depth knowledge of the region had been formed before the outbreak of war, when he acted as photographer and assistant to the archaeologists D.G. Hogarth and Sir Leonard Woolley. This camera was made for Lawrence – better known as Lawrence of Arabia – for use on an archaeological expedition to the ancient Hittite settlement of Carchemish in Syria. While on expedition he travelled throughout the region, producing detailed maps which became strategically significant when Ottoman forces allied with the Germans against the British. Lawrence's expertise in Arab affairs made him a valuable asset to the British cause when he joined the British Expeditionary Force in 1914.

The camera's compact size and interchangeable lenses ensured its versatility and portability and made it ideally suited for rough terrain and demanding environmental conditions. Quarter-plate hand cameras like this were popular with amateur photographers. The adjustable lens panel was particularly suited to correcting the perspective distortion of converging verticals in architectural and archaeological photography. Taken as a whole, Lawrence had acquired a well-appointed outfit for a serious photographer. The camera kit features a variety of lenses, including a wide-focus lens and telephoto lens. He certainly boasted in a letter home while on board ship to Syria in December 1910 that one of the lenses performed better than his binoculars in distinguishing features on the shoreline.

A simple handwritten inscription on the exterior of the sturdy leather carrying case reads: 'Property of T.E. Lawrence, Pole Hill, Chingford, Essex'.

Son, Scientist, Soldier

Moseley's X-ray spectometer

Manchester, 1913

Inv. 17217

Amabel Moseley's Collins Pocket Diary, 1915

MS Ludlow-Hewitt 1/15

H.G.J. 'Harry' Moseley (1887–1915) was an exceptionally promising young English physicist in the years immediately before the First World War. His work on the X-ray spectra of the elements provided a new foundation for the periodic table. Yet his life and career were cut short when he was killed in action at Gallipoli, Turkey, in August 1915, aged 27.

During a scientific research career spanning a mere forty months, Moseley worked under Ernest Rutherford (1871–1937) at Manchester for three years before returning to his Alma Mater, Oxford, in November 1913. Spanning his time at Manchester and Oxford, he developed an experimental set-up using ingenious apparatus of his own design to significantly change the world of science. Using X-ray spectroscopy, he discovered that every chemical element's identity is uniquely determined by its atomic number, representing the positive charge on its nucleus. Moseley provided a new order to, and understanding of, the periodic table of chemical elements and contributed to the development of the nuclear model of the atom.

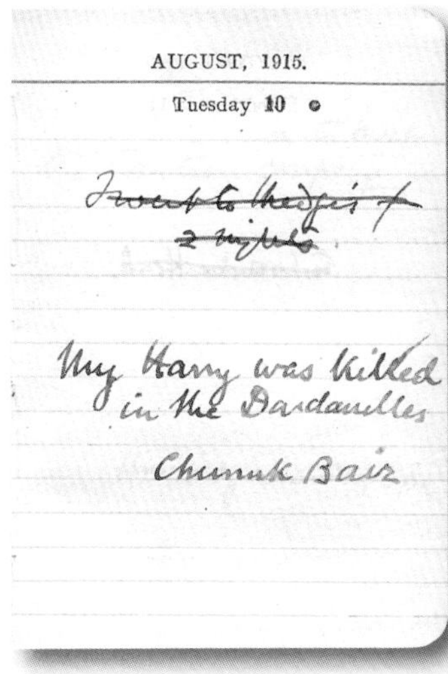

In October 1914 Moseley enlisted in the Royal Engineers as a signals officer. Meanwhile, his experimental apparatus, hand-drawn technical diagrams and related documents and ephemera were preserved by Professor Townsend at the Oxford Department of Physics and transferred to the Museum in 1935. After Moseley's tragic death in the Gallipoli campaign, his mother, Amabel, used her social diary to record her final moments with her son as well as the details of his death. She also worked with Ernest Rutherford, Moseley's former supervisor at the University of Manchester, gathering material now in the Museum's archives in order to establish and preserve Moseley's scientific and personal legacy.

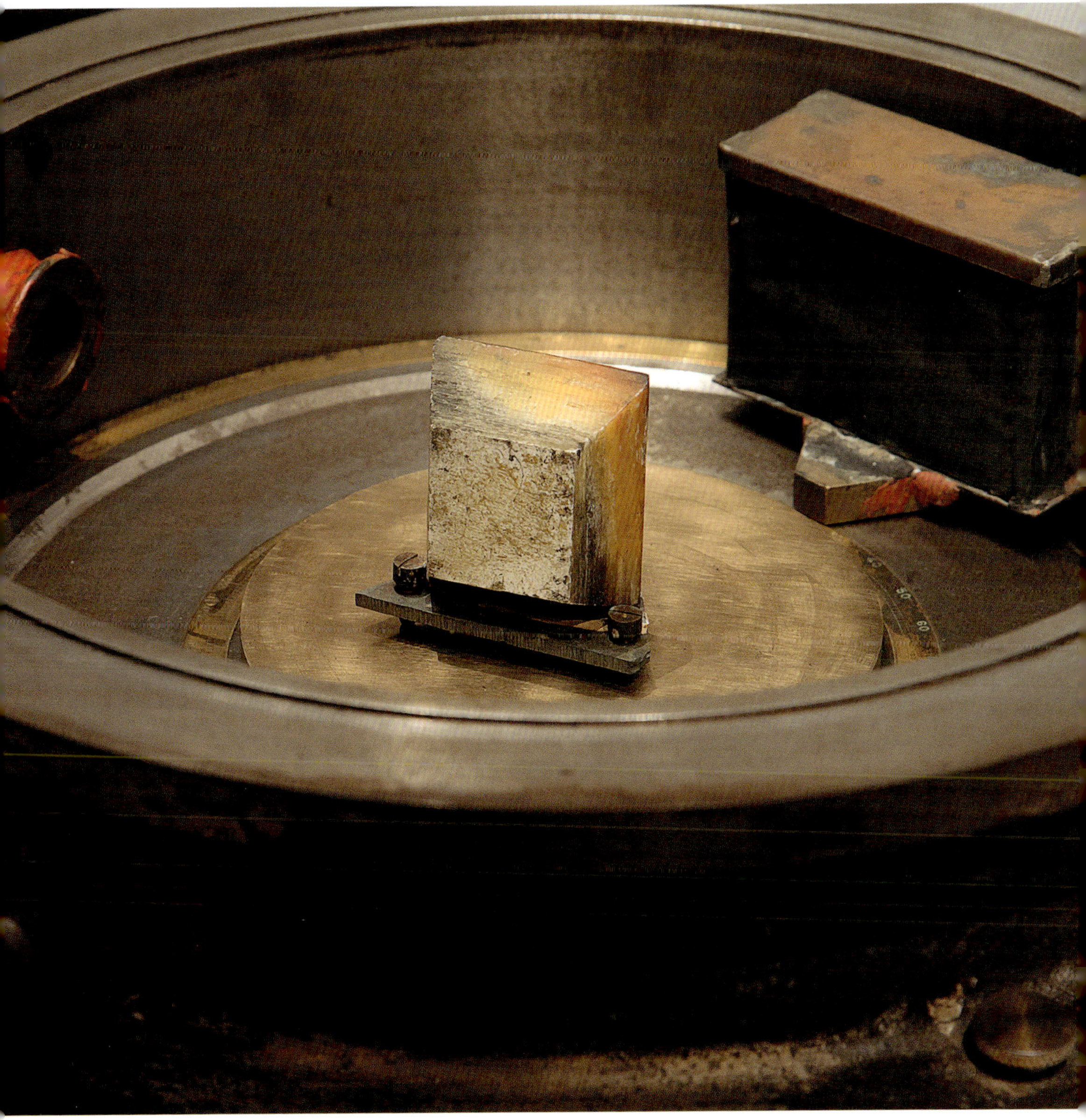

Art and Science Combined

Compound and simple microscope for George III

George Adams
London, *c.*1763
Inv. 35086

This magnificent silver microscope was made for King George III at the beginning of his reign. It has a fluted Corinthian column as its main support and the architectural ornament is enhanced by miniature vases on the base. Topped by two figures holding the barrel of the compound microscope, it also includes a simple microscope.

But was this fashionable magnificence or ridiculous extravagance? Does all the elaborate and ornate decoration help the instrument's function as a microscope? Not at all, from a technical point of view. Does being made in silver improve its performance? No, though it has required more conservation than brass usually demands.

Yet this was a working device, used not only for the private instruction and entertainment of the King and his children, but also accessible to visitors and widely reported as a symbol of the new King's enlightened taste. In that sense the microscope was successful and was already famous in its day. The leading French astronomer and mathematician Jérôme Lalande (1732–1807) paid it a special visit in 1763 and his diary makes clear that it was seen as more than just an expensive plaything:

> We went to see the king's microscope, which cost 3,000 *pièces* made by Mr Adam. It has all the lenses on a revolving disk and each one has its place; the mounting is of silver. It has a pointer between the oculars which distinguishes movement of a ten-thousandth of an inch to measure magnification.

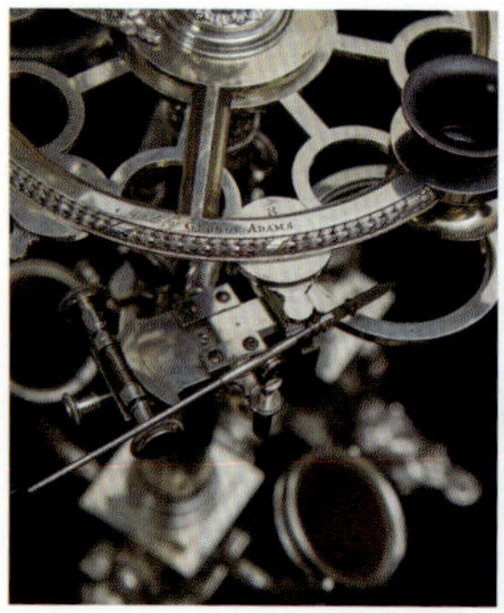

Of course, as well as advertising the King as a patron of the arts and sciences, it also enabled the royal instrument maker George Adams (1709–1773) to cultivate the fashionable interest in science and encourage customers to aspire to his most prestigious products.

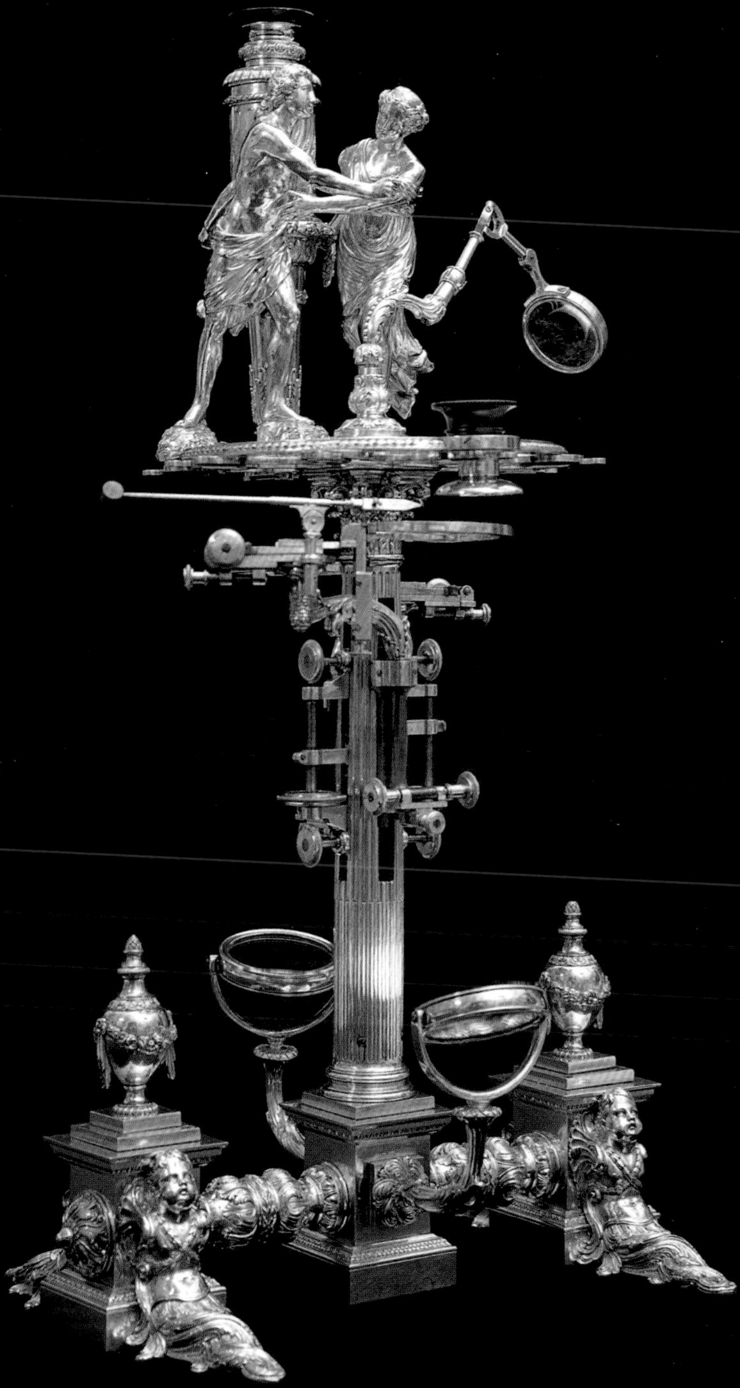

Gunpowder, Treason and Ptolemy

Armillary sphere

Italy?, *c.*1580

Inv. 70229

THIS LARGE and impressive brass armillary sphere is both a working astronomical instrument and a decorative object of display, with a finely modelled sculptural base. It is constructed according to the ancient geocentric view of the universe, as codified by Ptolemy in the second century CE. Surrounding the earth are the great circles of the heavens, with the sun in the broad ecliptic band of zodiac signs. Once set up for the user's latitude, it can solve problems of spherical astronomy, such as the time of sunrise on any day of the year.

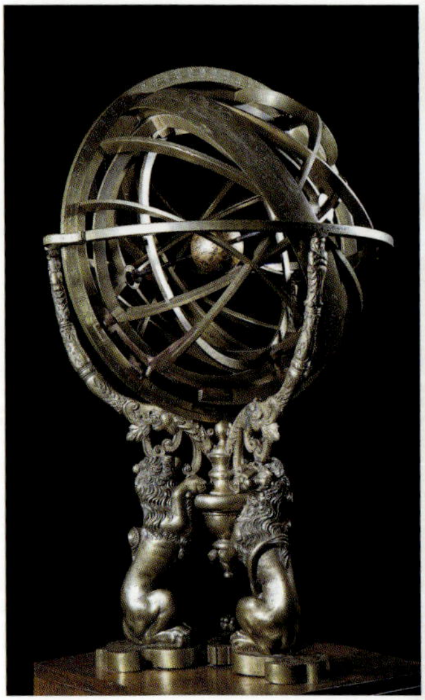

The instrument is unsigned and of uncertain origin, but is thought to have been made on the Continent. However, the heraldic devices engraved on the base show that it had arrived in England by the 1590s and come into the possession of Henry Percy, 9th Earl of Northumberland (1564–1632). Percy is known as 'the Wizard Earl' because of his interests in astronomy, astrology, alchemy and other branches of natural philosophy. He employed the leading mathematicians of his day, and was even visited by them during his 16-year imprisonment in the Tower of London on suspicion of involvement with the Gunpowder Plot of 1605.

Before this personal disaster, the sphere had already passed to Sir Josias Bodley (1550–1618), the younger brother of Sir Thomas Bodley (1545–1613), who founded Oxford's Bodleian Library. In 1601 Sir Josias presented the sphere to Bodley's library but it was kept at Sir Thomas's London house until his death in 1613. It has been in Oxford ever since, firstly on display in the Bodleian, then lent to the University Observatory in the late nineteenth century, and finally coming to the Museum of the History of Science in the 1950s.

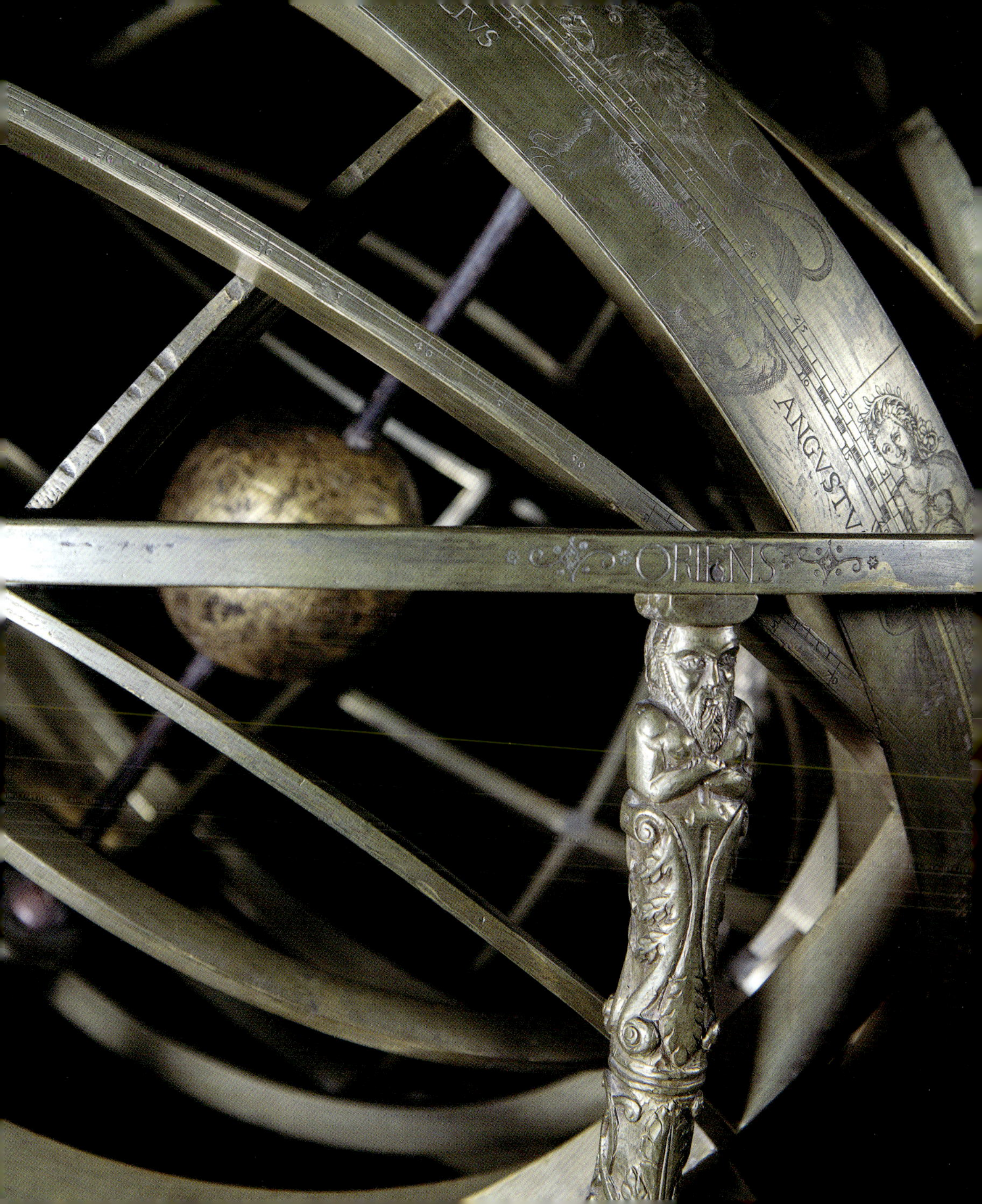
ORIENS
ANGVSTVS

Master of the Universe

Spherical astrolabe

MUSA
Eastern Islamic, 885 AH 1480/1 CE
Inv. 49687

THE UNIVERSE in your grasp. That is the promise of this spherical astrolabe, which is perfectly sized to nestle neatly in the palm of a hand. More than merely an image or a representation, it is a working device that captures and even seems to control the heavens.

All the cultures that inherited and developed Greek mathematics and philosophy agreed on the form of the world. The earth was at the centre of everything, orbited by the planets from the moon through to Saturn, and including the sun as the most splendid of these seven wandering bodies. Beyond the realm of the planets were the much more regular fixed stars, imagined as adorning an all-encompassing sphere that spun round once a day. What was beyond the stars? What moved the stars? That was the realm of God, the first or prime mover, ultimately responsible for all heavenly and earthly change.

To hold a spherical astrolabe was therefore not just to have a 'God's-eye view' of the world, but to act as the prime mover in miniature. The finely formed skeletal framework (the rete) that surrounds the inner sphere is the instrument's celestial component. It carries pointers for bright stars in the sky, so turning the rete round once represents a day. The rete's largest silver circle is the ecliptic, divided into the signs of the zodiac; on this the sun can be located for any day of the year. Rotating the rete thus moves the sun across the sky, enabling the astrolabe to solve many practical problems of time-telling for any latitude on earth.

If this seems an extraordinary and potent object, it is all the more so for being the only surviving complete example in the world.

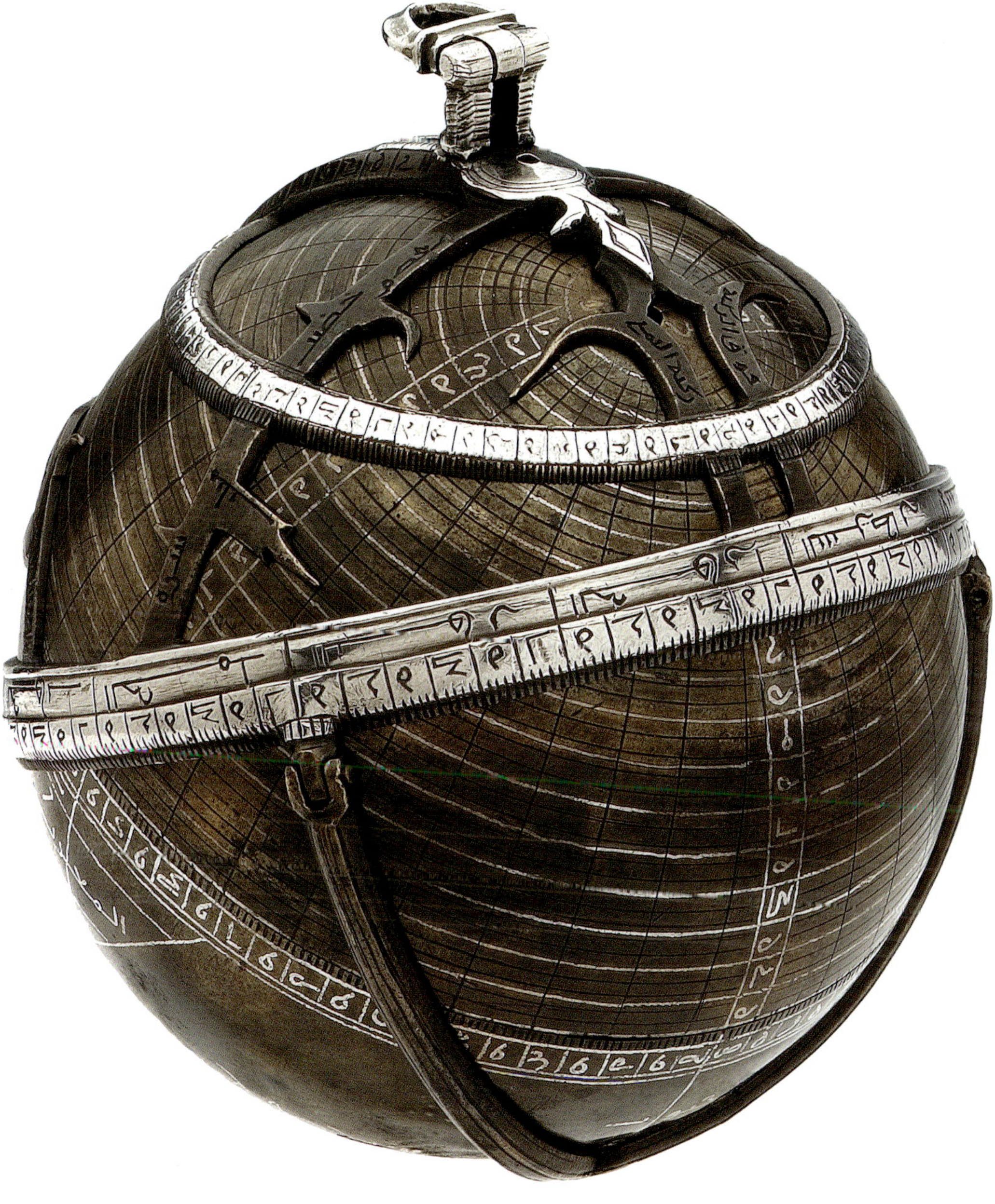

The Path of the Moon Engraved

Astrolabe with lunar mansions

'Abd al-Karim
The Jazira, probably Mayyafarikin, 625 AH 1227/8 CE
Inv. 37148

Every instrument made before industrialised manufacture was by definition unique. But this astrolabe probably has a stronger claim to this epithet than most: it is so far the only instrument known to display a figural representation of the lunar mansions.

Around the rim on the back of this astrolabe are the familiar images of the signs of the zodiac, given twice – in mirror image – to show the view from inside and outside the universe. The zodiac represents the star constellations through which the sun passes in its annual cycle. Nearer the centre is another band of figures. These represent the 28 lunar mansions (or houses). They were widely used in forecasting amongst the Arab Bedouins and continued to be of great importance after the advent of Islam in the seventh century CE. Like the zodiac signs, the lunar mansions are constellations of stars, but they derive from pre-Islamic folk astronomy. The meanings of their names had already been lost by the eighth century CE, making it extremely difficult to draw them. This silver-inlaid brass astrolabe is not only unique in depicting the lunar mansions, but its figures differ from those in the few manuscripts that also rose to the challenge.

'Abd al-Karim, the maker of this stunning astrolabe, is rightly famous for creating instruments that still dazzle us today with their intricate beauty and phenomenal craftsmanship. The inscription tells us that this one was made under royal patronage for al-Malik al-Ashraf Musa, one of the most powerful Ayyubid rulers of the thirteenth century.

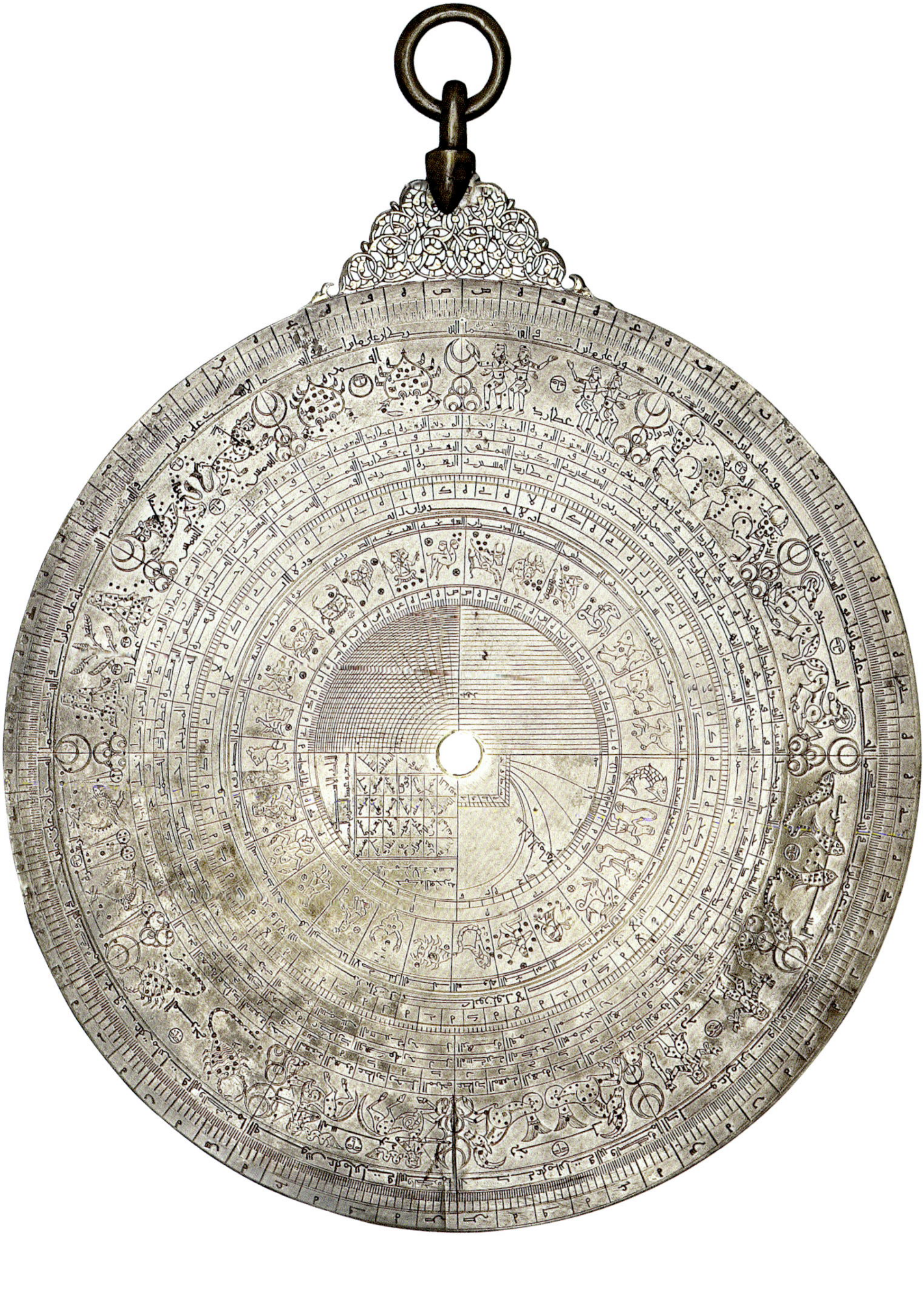

Founding Benefactor's Choice

Diptych sundial

TUCHER WORKSHOP
Nuremberg, *c.*1600
Inv. 41986

Portrait of Lewis Evans

W.E. MILLER
England, *c.*1920
Inv. 62096

THE IVORY DIPTYCH dial was the quintessential Renaissance gadget. To call it a sundial is to underplay its ambition and ingenuity – it was the sixteenth-century equivalent of a smartphone with a multitude of apps. With a compass to orient the instrument correctly, the device opened up to reveal multiple sundials of different types, telling the time in different hour systems and for use in different places.

Added bonuses typically included an indication of where the sun was in the zodiac and a rotating disc or volvelle to provide calendrical, astrological and lunar information. There was even sometimes a wind vane for weather prediction. Diptych dials were usually uniquely designed and manufactured, with neat and colourful decoration to set an owner's stamp of pride on a stylish object of desire. Nuremberg was the European centre of manufacture but the raw material came from African elephant tusks, so the finished products were part of a global system of trade.

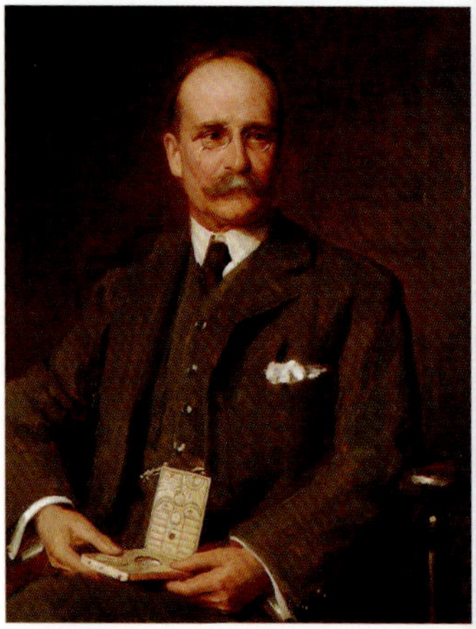

The instrument here is a fine but fairly typical example, dating from about 1600. Why select it from the literally dozens in the Museum's collection? Because it appears in a portrait of Lewis Evans (1853–1930), the founding benefactor of the Museum.

Evans was passionate about mathematical instruments, particularly those used for timetelling. He was scholarly in his interests, assembling a significant mathematical library to illuminate the objects he acquired, which were all individually catalogued on cards. His whole collection of thousands of instruments and books was offered to the University of Oxford, where it was initially displayed in the upper gallery of the current Museum and known as the Lewis Evans Collection.

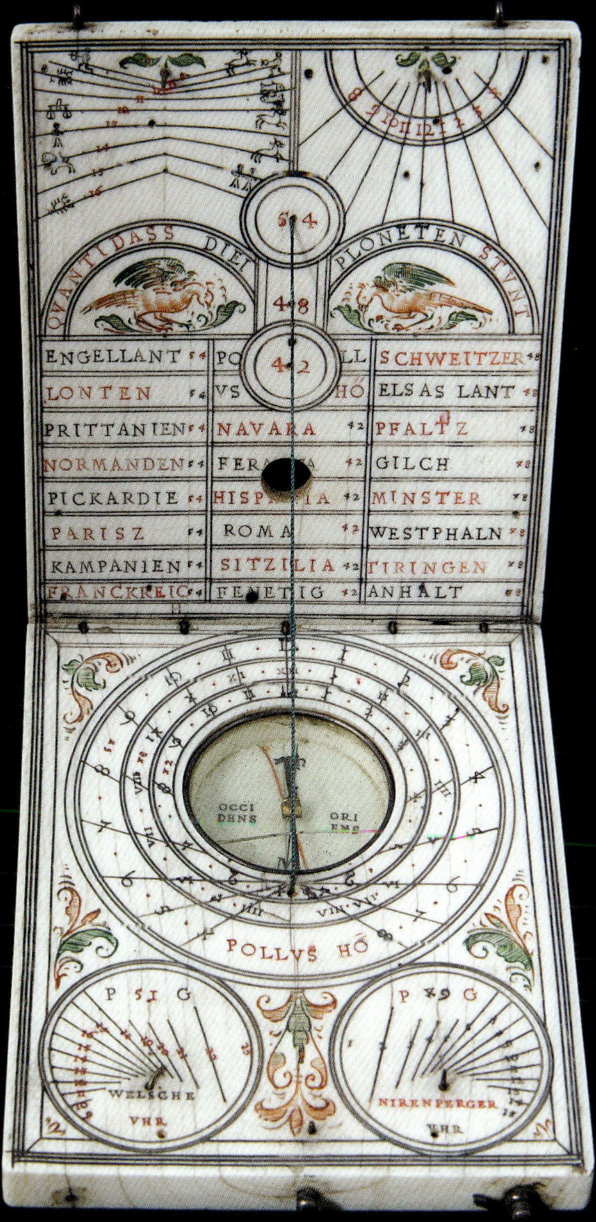
QVANTI DASS DIEI
PLONETEN STVNT
ENGELLANT
SCHWEITZER
LONTEN
ELSAS LANT
PRITTANIEN
NAVARA
NORMANDEN
GILCH
PICKARDIE
MINSTER
PARISZ
ROMA
WESTPHALN
KAMPANIEN
SITZILIA
TIRINGEN
ANHALT
OCCI DENS
ORI ENS
POLLVS HÖ
WELSCHE VHR
NIRENPERGER VHR

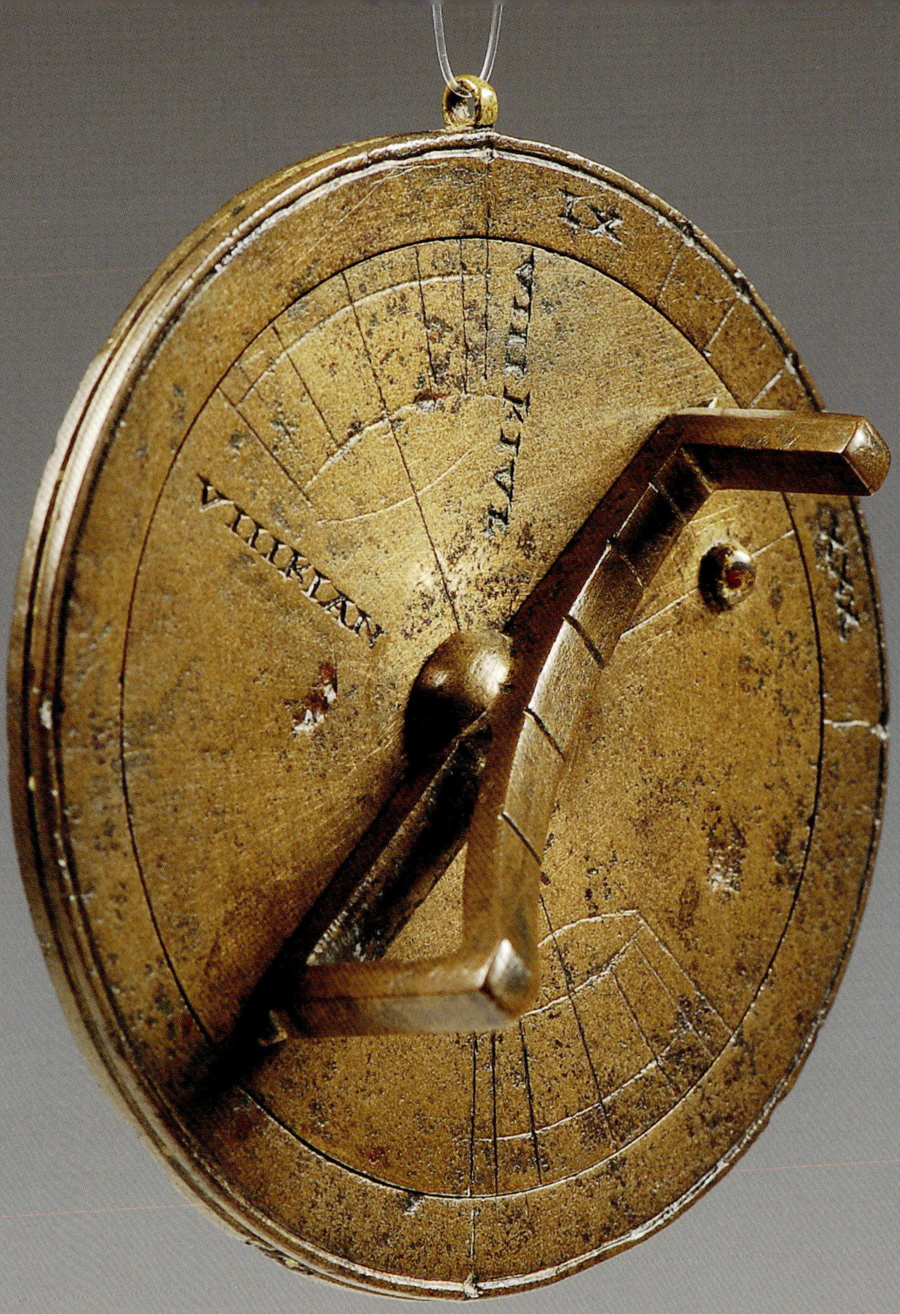

Time in Your Hand

Roman vertical disc sundial

Found near Bratislava, *c.*250 CE

Inv. 51358

In a world before mechanical clocks, how did people tell the time? Water clocks were one popular method and sundials another. Many sundials remain from classical Greek and Roman antiquity, but almost all were fixed in one place: carved in stone and meant to serve as public or private monuments. This tiny sundial is one of just a handful of ancient portable dials to survive. Though only 6 cm across, it mirrors the whole Roman world.

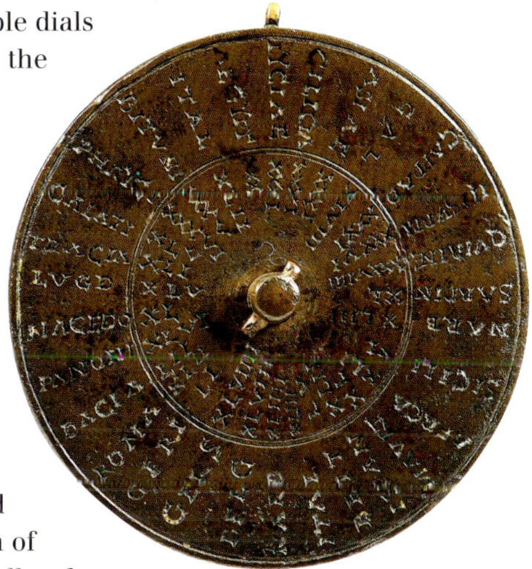

As a portable dial, it has to work in a range of latitudes. To ensure it is correctly configured for the owner's location, there is a circular table on the back giving the latitudes of thirty provinces of the Roman Empire. After setting the latitude, the dial is adjusted for the current date. Then it is suspended and turned until the horizontal bar of the gnomon or pointer casts its shadow on the curved hour-scale below, thereby indicating the time.

Most sundials are geometrically exact and their arrangement faithfully mimics the motion of the sun in the sky. Even if they are in practice small and imprecise, at least in principle they will always give correct solar time. This dial is different. It is based on an approximation, which, although generally accurate, means that the heavens are not exactly captured in the layout of the instrument. In a society far removed from our modern preoccupation with minutes and seconds, this would not have been a practical problem. But it does reinforce the common stereotype that the Romans were concerned more with results than with theoretical rigour.

Experimental Philosophy in Public

Air pump

Francis Hauksbee the Younger
London, *c.*1715
Inv. 60664

The air pump is one of the most iconic scientific instruments. A product of the ferment of new 'natural philosophy' in the middle of the seventeenth century, it is now an emblem of the new methods of understanding nature, and even of the emerging values and culture of modern science itself.

Oxford has a particular connection with the development of the device: it was here that Robert Boyle (1627–1691) and Robert Hooke (1635–1703) made an air pump and studied the remarkable effects inside the artificial vacuum it produced. But almost nothing survives of the earliest machines. Their glass receivers broke, they were cannibalised for improvements and were never intended as prestigious, collectable devices.

In the years after 1700 the air pump moved from private research to public lectures and fashionable demonstrations. It is from this period that instruments survive. This particular example – which may have been used in the Museum building in the eighteenth century – exactly follows the pattern established by the most notable expert of the time. Francis Hauskbee the Elder (1660–1713) was appointed by Isaac Newton (1643–1727) to be the demonstrator of the Royal Society in 1703. However, he first comes into view without such illustrious associations. He had already improved the practice of cupping (bloodletting) and it was his technical innovations in the manipulation of air that eventually led to the classic form of the double-barrelled air pump. Surprisingly, his early advertisements link him to the 'hummums' of London – Turkish baths where medical and more disreputable services were on offer. From such unlikely origins, a reliable, rapid and rugged device resulted, which became the standard for decades to come. This example, perhaps made by Hauksbee's nephew (1687–1763), was still in use in the later eighteenth century, as witnessed by a plate marking its repair.

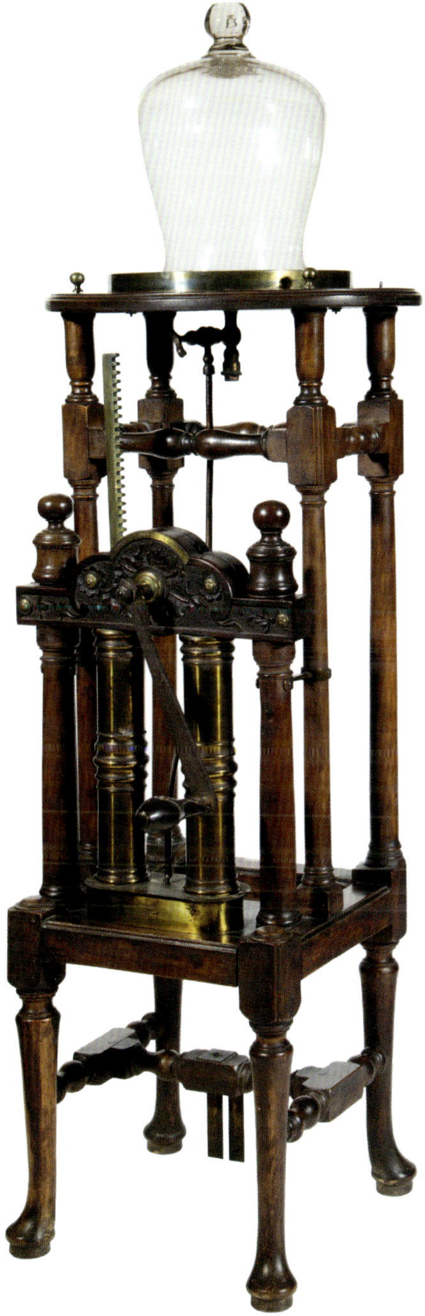

The Great Moon Hoax

Snuffbox depicting Sir John Herschel

Germany, *c.* 1835

Inv. 25152

On permanent loan from the Royal Astronomical Society

A STRANGE ASSORTMENT of gentlemen stand over, or actively peer into, the image formed by an enormous optical apparatus. The caption, in both German and French, reads 'Dr John Herschel's newest discoveries in the world of the moon'. Herschel (1792–1871) was a leading astronomer, then based at the Cape of Good Hope. But this small snuffbox and its astronomical image had their origins neither in Europe nor in Southern Africa. Rather, its story begins in New York with a cheap daily newspaper, *The Sun*, whose front page for Tuesday 25 August 1835 carried the headline 'GREAT ASTRONOMICAL DISCOVERIES LATELY MADE BY SIR JOHN HERSCHEL, L.L.D. F.R.S. &c.' In this and the next five editions the sensational news emerged that Herschel's giant telescope had observed life on the moon. The paper's circulation shot up: everyone wanted to read for themselves the extraordinary revelations of bizarre plants, beavers that walked upright, bat-men and temples.

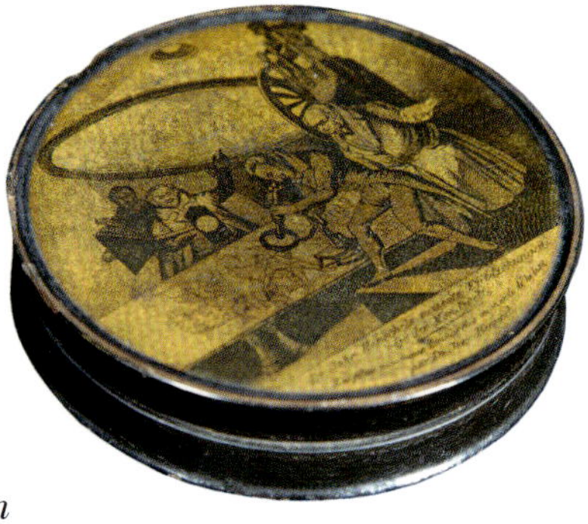

Some were sceptical from the first. Others were unsure but appreciated that, even if it was no more than a modern version of *Gulliver's Travels* or *Robinson Crusoe*, the illusion had been created with convincing skill and detail. However, by the end of the serialisation the game was up and it was being treated more widely as a hoax.

Yet it did not die. Pamphlets reprinted the material in Britain, France, Germany, Italy and Spain, often with illustrations. Maybe some who eagerly read the story were doing so more to laugh at the gullibility of those who had initially been taken in. But whatever the motivation, this snuffbox is part of a global phenomenon, as news and newspapers impacted the whole world in the early nineteenth century.

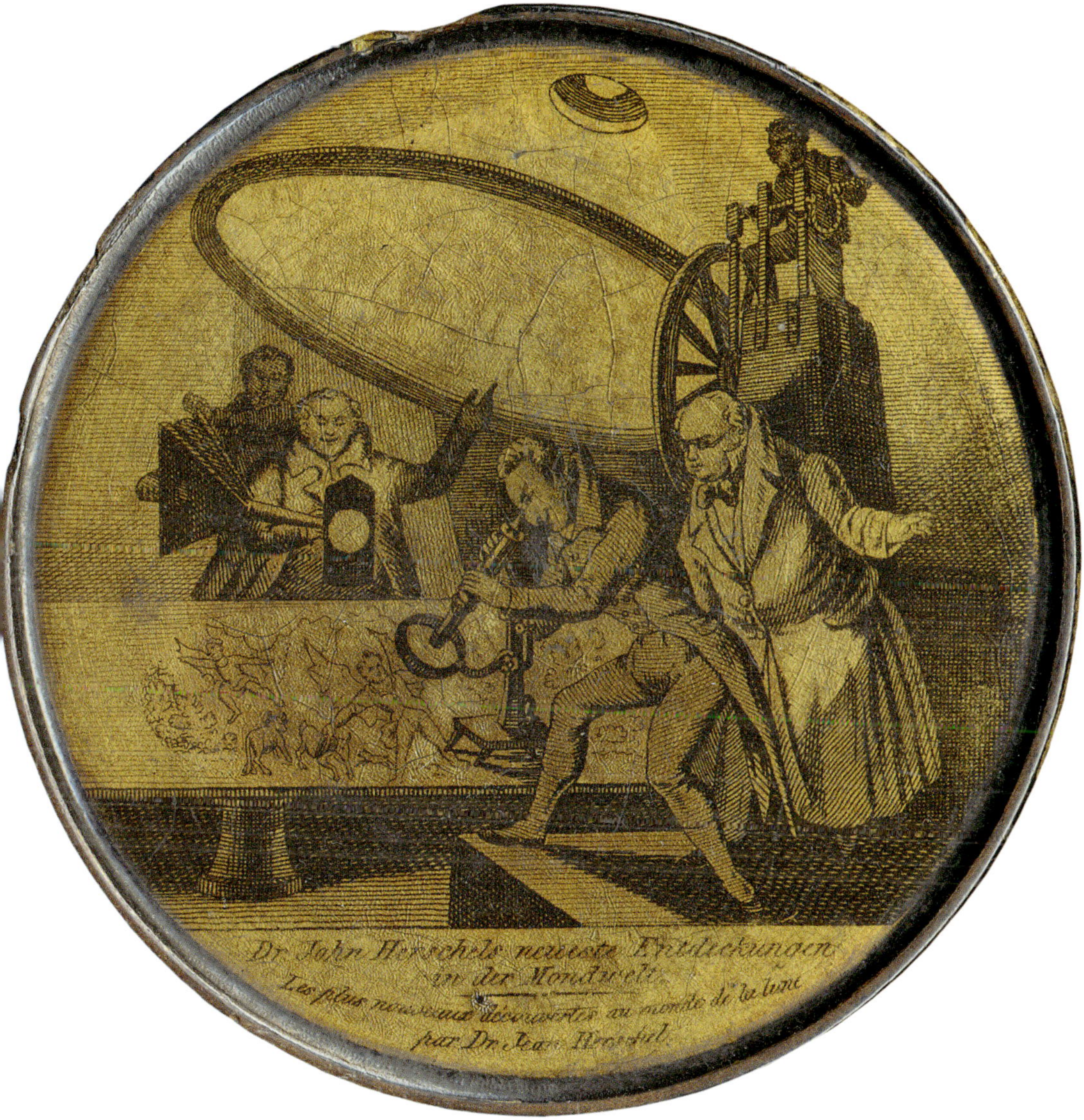
Dr John Herschels neueste Entdeckungen
in der Mondwelt
Les plus nouveaux découvertes au monde de la lune
par Dr Jean Herschel.

From Scientific Mystery to Military Medicine

Jackson-type X-ray tube

Newton & Co.
London, 1898
Inv. 69823

Wimshurst twin plate electrostatic machine and fluorescent screen X-ray accessory

Harvey & Peak
London, 1898
Invs 82113, 78574

When Wilhelm Röentgen (1845–1923) announced his discovery of X-rays in 1895 he caused a scientific sensation. That it was possible to see inside a living body provoked amazement amongst the wider public too. But how these new rays worked, and exactly what they were, was deeply mysterious – hence their name, X being the unknown.

This group of objects demonstrates the enthusiasm for the new branch of physics, even outside the laboratory. The apparatus also shows how quickly the fashionable excitement around X-rays could be transformed from curiosity to powerful uses.

There are three items here, which were purchased and used together by an Oxford student in 1898. Edward George Spencer-Churchill (1876–1964) was a first cousin of Winston Churchill and an undergraduate at Magdalen College when he decided to try out the still-novel phenomenon of X-rays. He acquired an X-ray tube and a Wimshurst machine, whose rotating discs generate the electricity delivered to the glass tube. Electrons from the tube's cathode strike the metal target of the anode to create the X-rays. The set was completed with a fluorescent screen.

What earnest scientific subject was Spencer-Churchill pursuing? He was primarily taking images of the limbs of his friends – for fun! But amusement soon gave way to more serious concerns. On the outbreak of the Boer War in 1899, he became a Lieutenant in the Grenadier Guards and the apparatus was taken to South Africa, where it was used to assist the Royal Army Medical Corps to locate bullets, shrapnel, splinters or fractures in wounded servicemen. It was in use there between 1900 and 1902, and was still unscathed when it was presented to the Museum in 1939 by Spencer-Churchill himself.

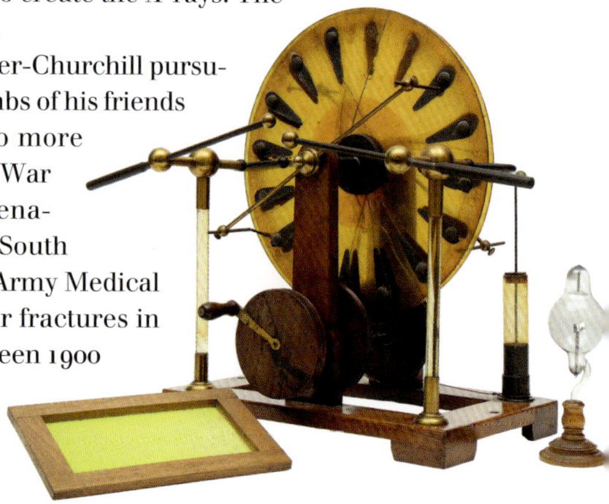

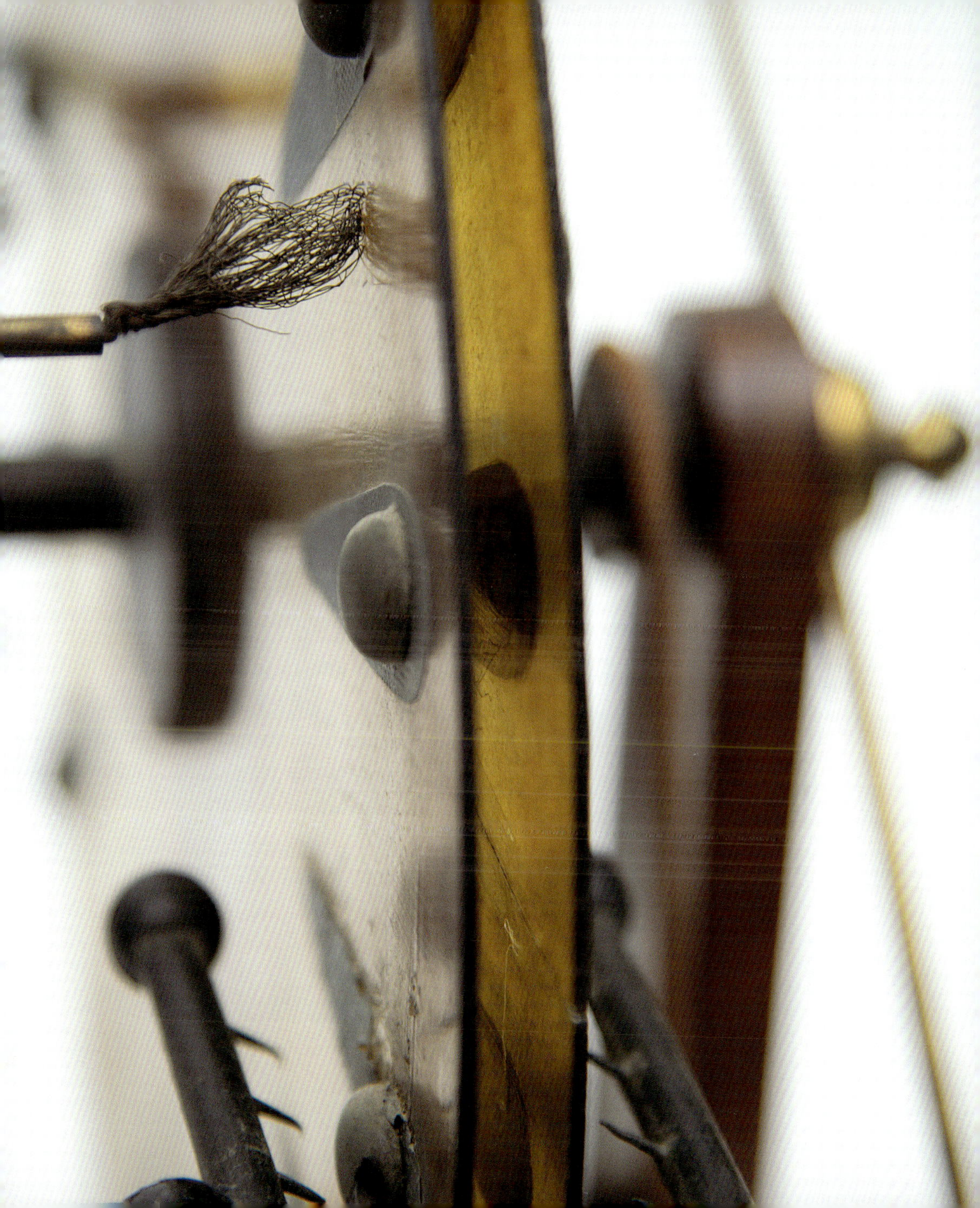

A Calculator for Hitler

Curta mechanical calculator Type I

CONTINA
Liechtenstein, *c.*1950
Inv. 34662

THE PINNACLE of mechanical computing miniaturisation was achieved by the fully featured Curta calculator, first introduced in 1947. The Curta is a small, finely made, hand-cranked digital mechanical calculator that performs multiplication and division as well as addition and subtraction.

Austrian engineer Curt Herzstark (1902–1988) designed this pocket calculator, initially conceiving the design in the 1930s. By 1938 he had filed a key patent for the unusual design. His work was temporarily halted when the Nazis forced Herzstark and his company to manufacture instruments for the German army. The son of a Jewish father and Catholic mother, he was taken into German custody in 1943 and was eventually sent to Buchenwald concentration camp. Here he was ordered to continue work on his mechanical calculator, both as a useful invention for German army engineers and also as a gift for Adolf Hitler to celebrate the expected German victory at the end of the war.

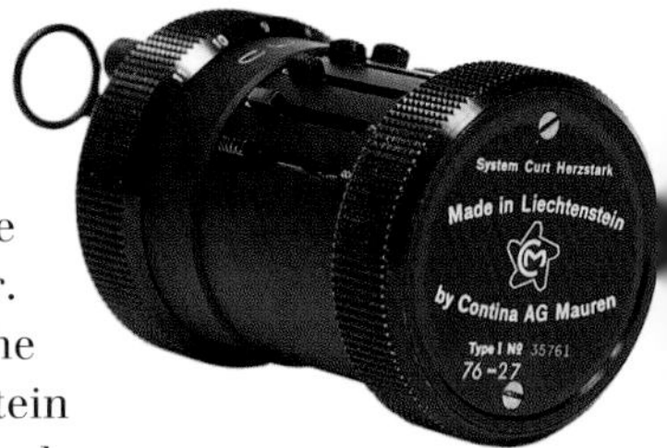

Liberated by Russian forces in 1945, Herzstark feared being shipped off to Russia so he fled to Austria, where he began to look for financial backers for his Curta calculator. The Prince of Liechtenstein eventually showed interest and the earliest Curta calculators were manufactured in Liechtenstein from April 1947 onwards. Approximately eighty thousand examples of the Type I were made before production ceased in November 1970. The machine remained the best portable calculator available until it was displaced by electronic calculators in the early 1970s.

This particular Curta calculator, an early Type I example made by Contina AG in Mauren, Liechtenstein, was formerly used by Dr A.J. Croft of Oxford's Clarendon Laboratory. In this physics laboratory it was known as 'Dr Croft's snuff-grinder'.

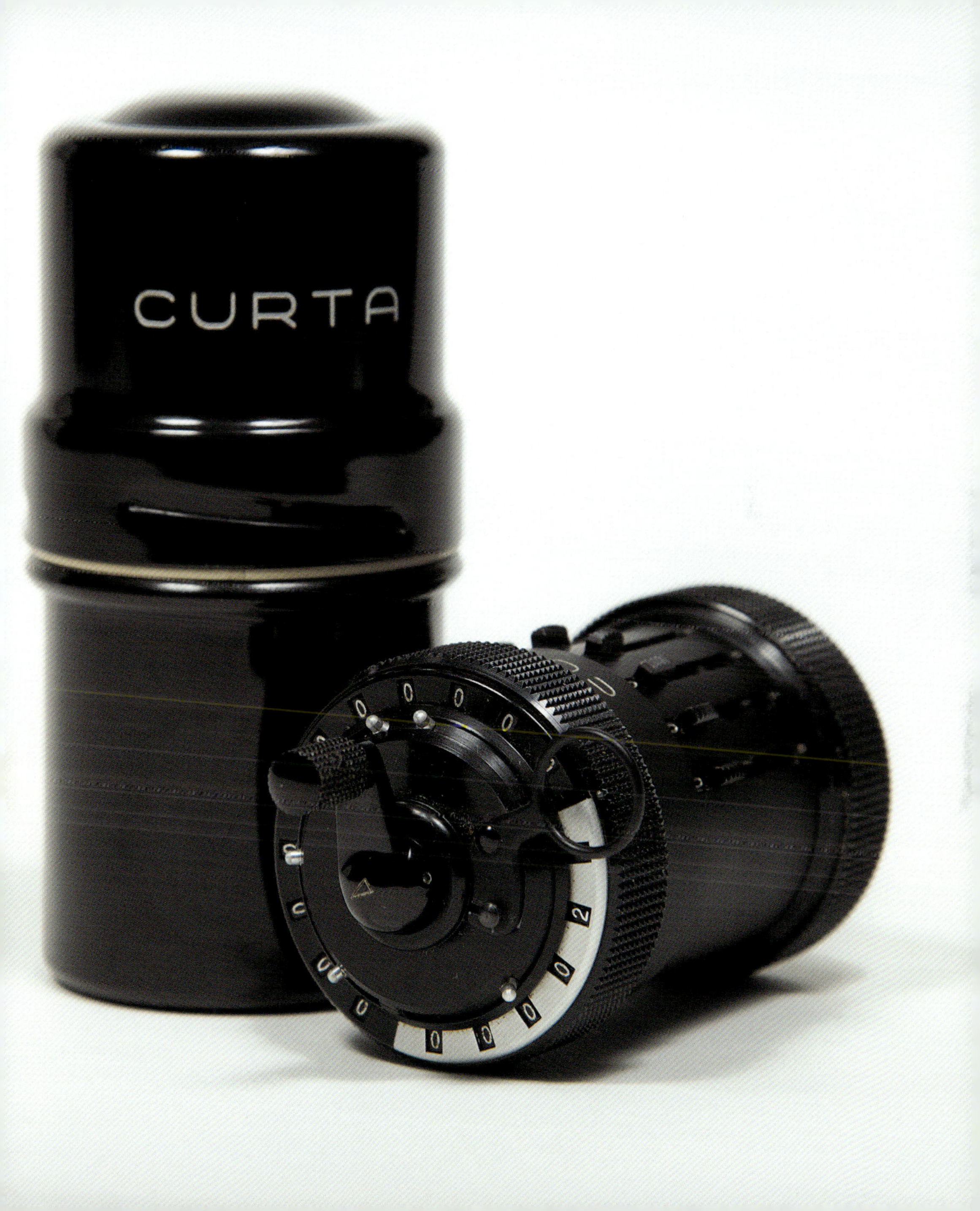
CURTA

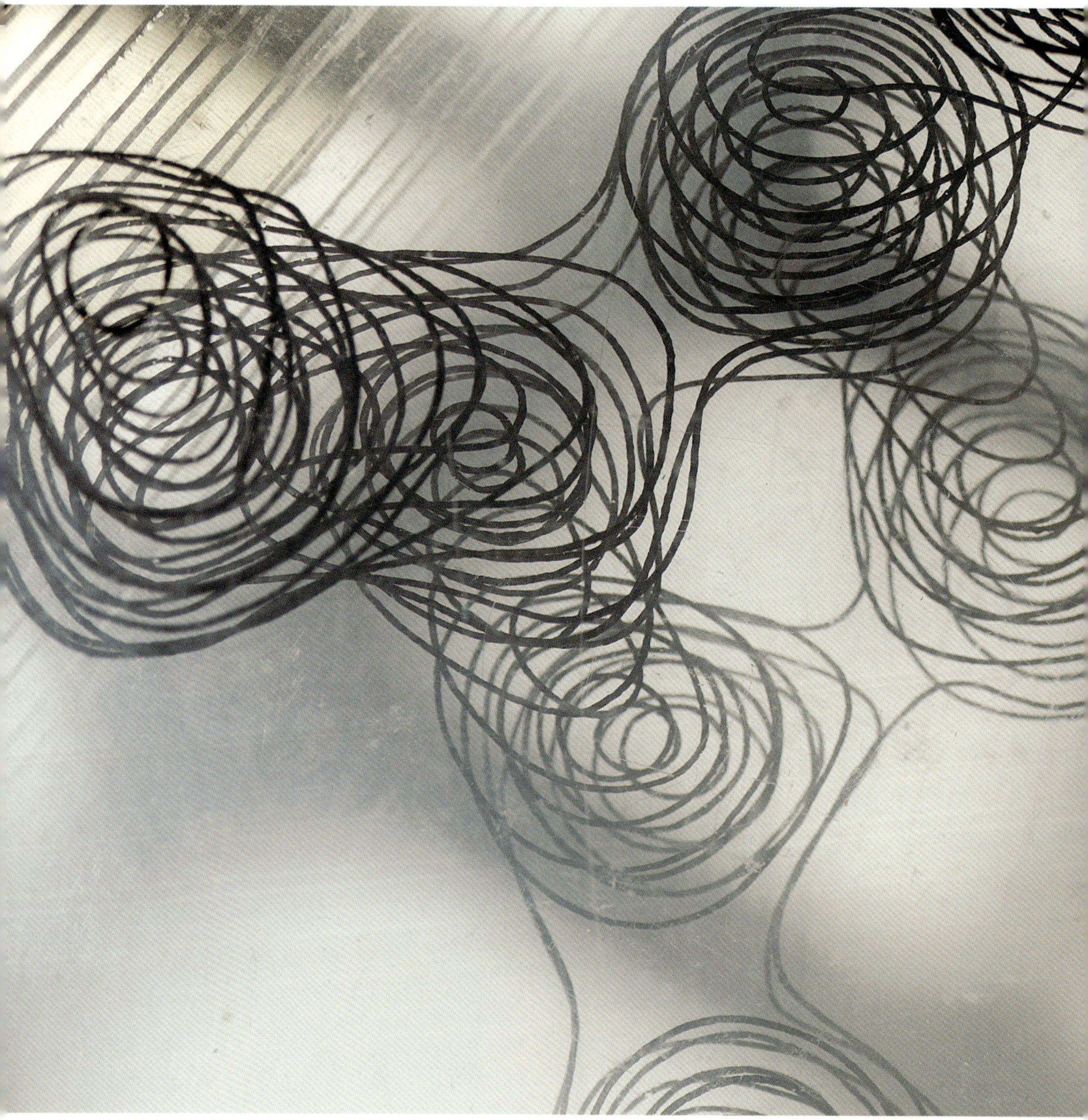

Mapping Penicillin

Model of the structure of penicillin

Dorothy Hodgkin
Oxford, *c.*1945
Inv. 17631

At the start of the Second World War penicillin was a major research focus at the University of Oxford. While a team at the Dunn School of Pathology pioneered its clinical use, the crystallographer Dorothy Hodgkin (1910–1994) took up the problem of establishing its molecular structure. Beginning in 1942, Hodgkin worked with several collaborators on the forbiddingly difficult task of extracting three-dimensional structures from the two-dimensional pattern of spots on X-ray photographs. In an age before digital computers, this involved painstaking processes of manual calculation, where many practical techniques were devised to cope with the complexity.

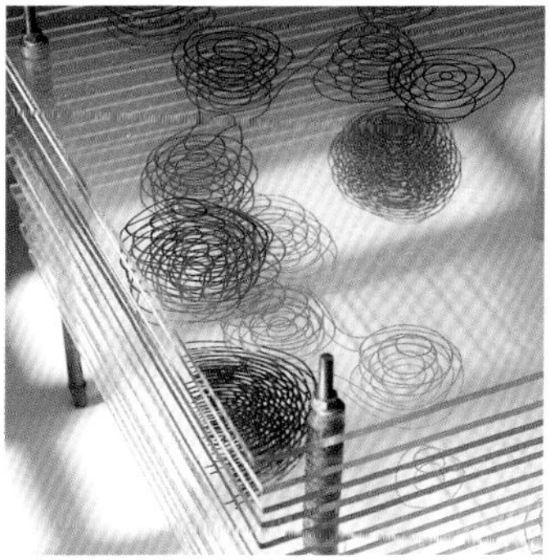

To achieve a three-dimensional picture, Hodgkin built up her model of the penicillin molecule in slices drawn on sheets of Perspex. The black contour lines represent electron density: the more concentrated the lines, the heavier the atom whose location was being revealed. By looking down through the layers of plastic it was possible to grasp the arrangement of individual atoms in the molecule. This artefact of Hodgkin's research is a rather evocative example of three-dimensional scientific modelling being used to image an invisible structure before the age of computer-driven virtual imaging.

After her work on penicillin Hodgkin researched a range of other biologically significant molecules and in 1964 was awarded the Nobel Prize in Chemistry. Underlying all this research was Hodgkin's ambition to establish the structure of insulin. In 1934, when insulin crystals were first given to Hodgkin, X-ray crystallography was unable to offer insight into the structure of such a complex molecule. Refining imaging techniques, she finally revealed the structure of insulin 35 years later, in 1969.

Mathematical Order

Architectural and geometrical model

England, *c.*1620

Inv. 90861

The alabaster 'mathematicall pillar' given to the University in 1620 by Sir Clement Edmondes (1564–1622) captures a peculiarly Renaissance enthusiasm for mathematical learning. The central pillar is a pentagonal column and each of its five sides displays one of the five classical orders of architecture, so that the complete series of Tuscan, Doric, Ionic, Corinthian and Composite is portrayed.

Architecture was then considered as one of the mathematical arts or sciences. But there is another famous group of five mathematical objects: the Platonic solids presented in the ancient text of Euclid's *Elements of Geometry*. Euclid proves that there can only be five of these regular solids, starting with the pyramid-like tetrahedron and continuing to the dodecahedron with pentagonal faces. Edmondes's model juxtaposes these regular solids with the architectural orders; unfortunately not all the skeletal solids have survived and those of brass are presumably replacements for the fragile alabaster originals.

Surmounting the whole is a dodecahedron, the only one of the Platonic solids that can symmetrically sit on top of a pentagonal column. It also happens to be Plato's symbol for the whole universe, giving this model a powerful resonance in the world of England's late-flowering Renaissance.

Edmondes's donation came shortly after the completion of the Schools Quadrangle (part of the Bodleian Library, founded by Sir Thomas Bodley. The quadrangle's centrepiece is the famous Tower of the Five Orders, which displays the five architectural orders one above the other. The model also echoed its location in another way: its donation exactly coincided with the founding of Oxford's Savilian professorships of geometry and astronomy by Sir Henry Savile (1549–1622). Savile's professors taught in the Bodleian's Tower and Quadrangle, so the model perfectly embodies the contemporary vitality of classical learning.

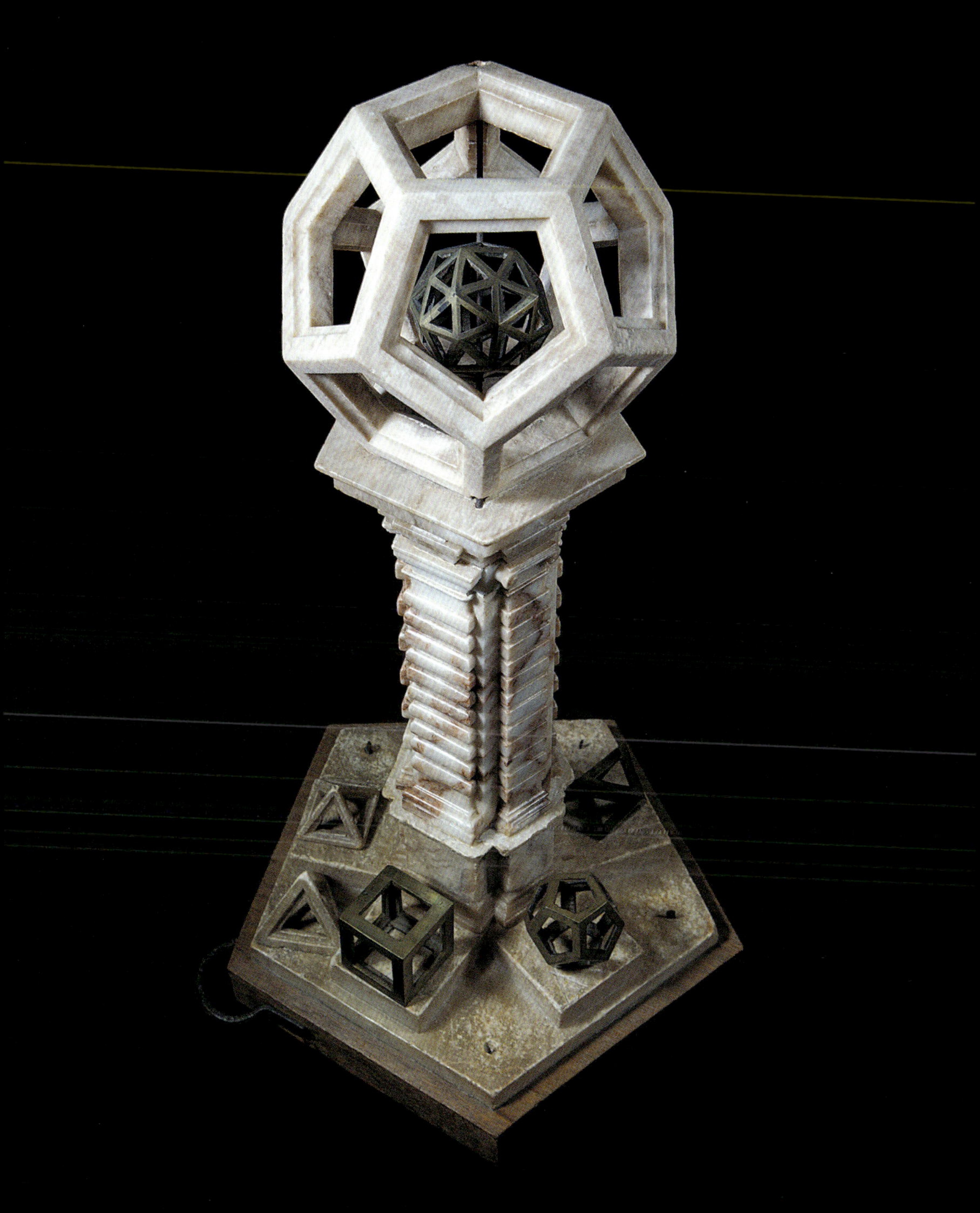

Enlightened Bird-Nesting

Collection of birds' eggs in wooden drawer and illustrated booklet

WILLIAM JONES
England, *c.*1799
Inv. 11117, MS Museum 281

A DRAWER long-since removed from a lost cabinet, sitting on a shelf in the Museum's store, contains a collection of birds' eggs. Collected in the 1790s, they are in remarkably good, clean condition for their age. Accompanying them is a small home-made booklet of coloured, life-size illustrations of eggs, dated 1799. Interestingly, it is not a catalogue of the collection; it is an unfinished attempt at a systematic iconography of British birds' eggs, as a basis for their classification. The collection itself contains 96 eggs, all of British birds, combining accessible mainland birds with an unusual number of sea-birds' eggs, some of which must have been taken from very remote locations. Sailors and fishermen probably brought them back on behalf of the collector, William Jones (1745–1818).

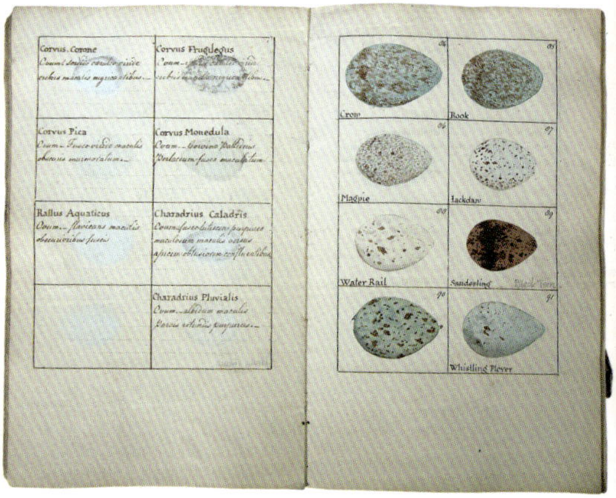

Jones was a wealthy London wine merchant, amateur naturalist and illustrator, celebrated even in his lifetime for his coloured drawings of British butterflies. After completing the butterfly iconography he turned to birds' eggs, but abandoned the project due to his failing eyesight. Much later, his material was given to the University of Oxford; the butterfly illustrations are held by the Museum of Natural History.

While scientific instruments lie at the heart of our Museum, the collection embodies a broader view of scientific heritage, hence the richness and variety of material culture represented in these pages. Several other collections or cabinets of natural history and materia medica are held, illustrating the importance of systematic collecting in the development of science.

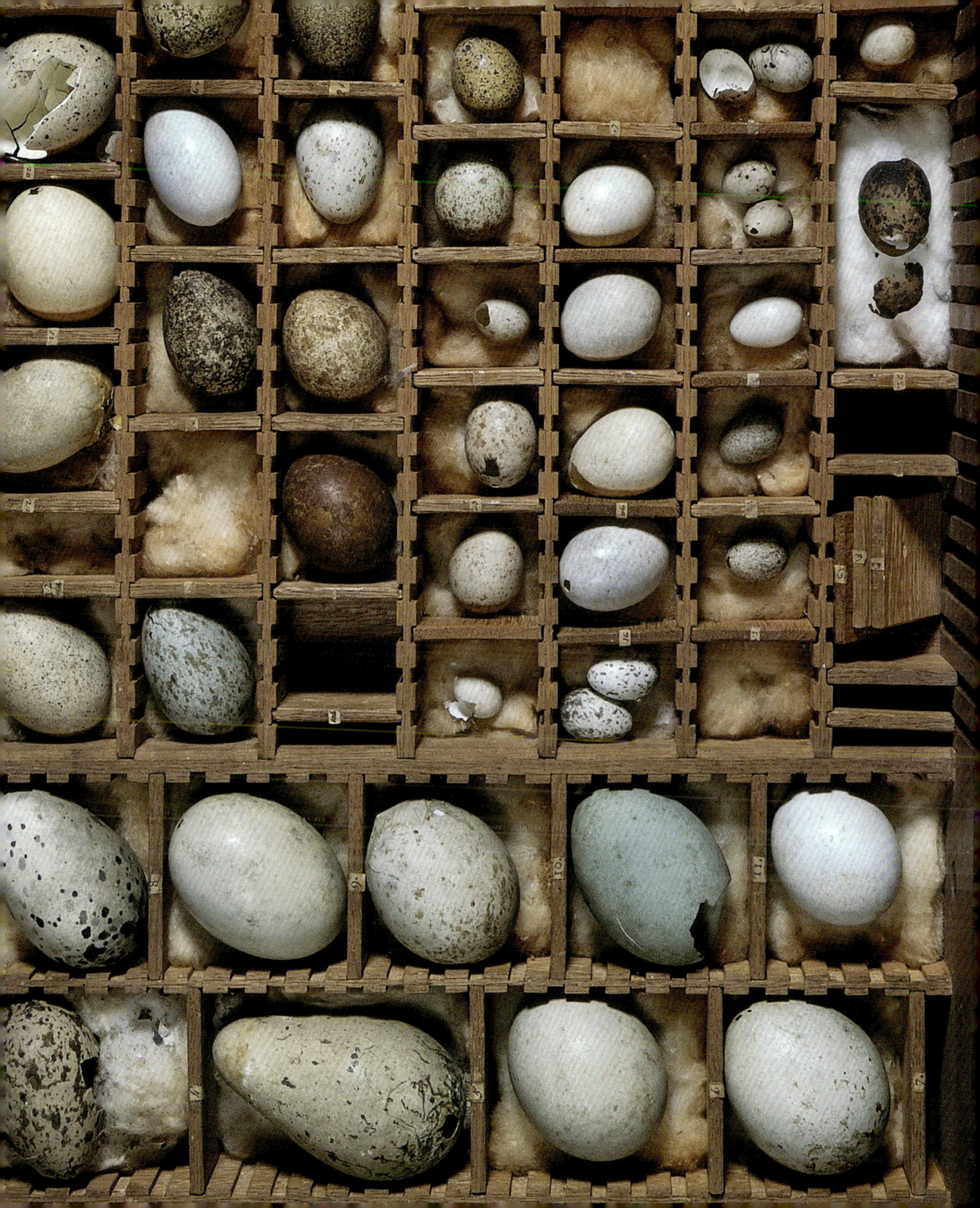

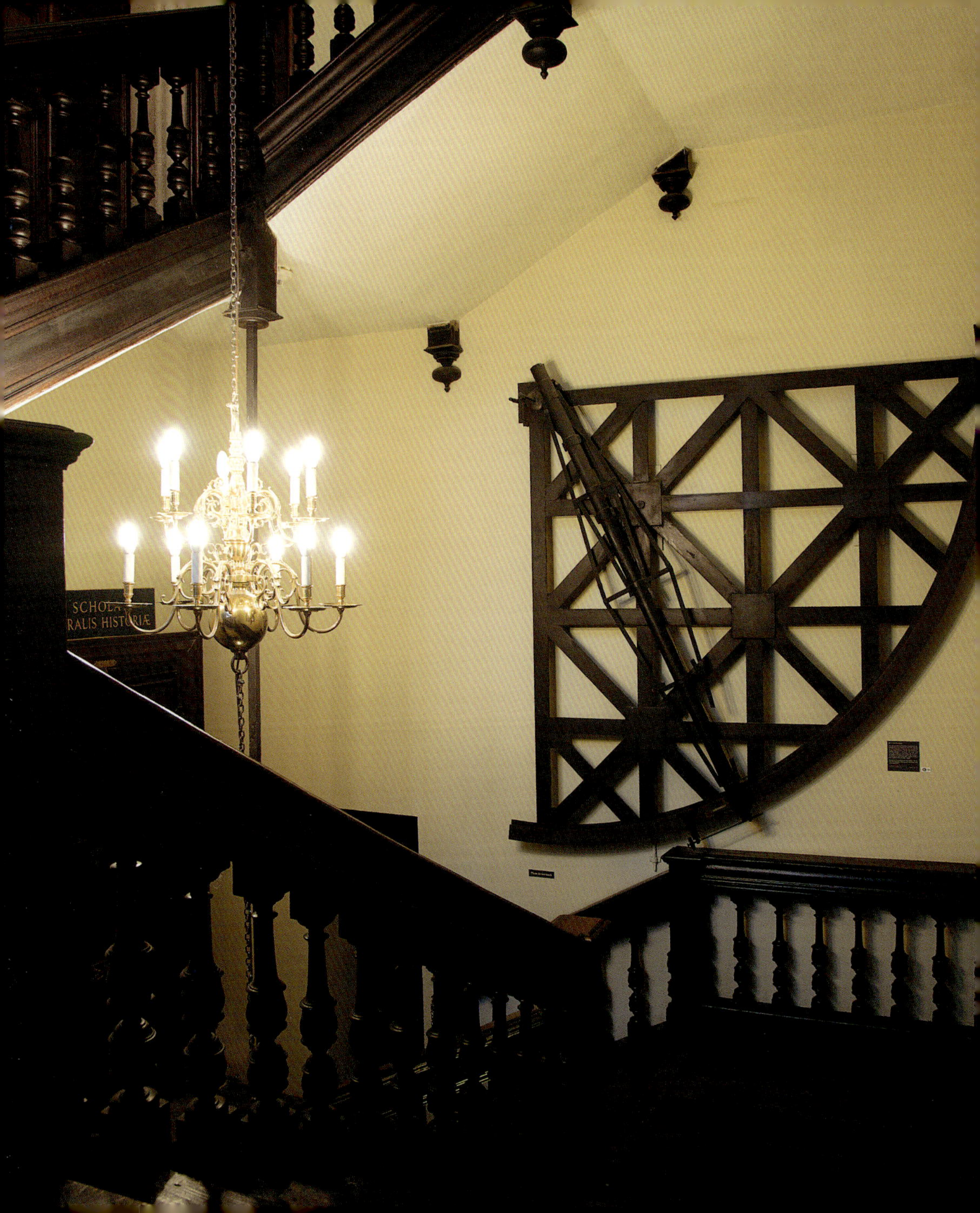
SCHOLA
RALIS HISTORIÆ

Astronomy Up Against The Wall

Eight-foot mural quadrant from the Radcliffe Observatory

John Bird
London, 1773
Inv. 30919

Many visitors are amazed, in a collection so rich in small hand-held and table-top devices, to see large astronomical instruments dominating the Museum's staircase. This brass 8-ft mural quadrant is one of a group made between 1772 and 1775 by John Bird (1709–1776) as the foundation equipment of Oxford's new Radcliffe Observatory. Bird supplied two of these mural quadrants, an 8-ft transit, a 12-ft zenith sector and an equatorial sector. They were permanently mounted instruments for measurement, not just for observation. All survived, along with numerous smaller and later instruments, when the Observatory closed in 1935, and the majority were transferred to the Museum of the History of Science.

The quadrant was mounted on a wall to point directly northwards in the meridian, with its twin pointing southwards in the opposite direction. Both were used only for measuring the height of stars above the horizon. Bird had established a reputation as the best maker of astronomical instruments in the world and his quadrants were famous. He had his portrait painted to show him with a published print for this design of quadrant.

Oxford's Savilian Professor of Astronomy Thomas Hornsby (1733–1810) was the prime mover behind the foundation of the new Radcliffe Observatory. It was funded by the Radcliffe Trust, which had already been responsible for prestigious new buildings in Oxford, such as the Radcliffe Camera. In his application, Hornsby emphasised that it was crucial to have the right instruments, concluding: 'There is unquestionably but one Person living, who is capable of making them'. He used the fact that Bird was by now in his sixties, and often in poor health, to pressurise the trustees into rapid action – in fact it was the last major project of Bird's life.

Von wegen bessers verstandts der vorgesetzten wort
habe ich alhie dise figur gemalet.

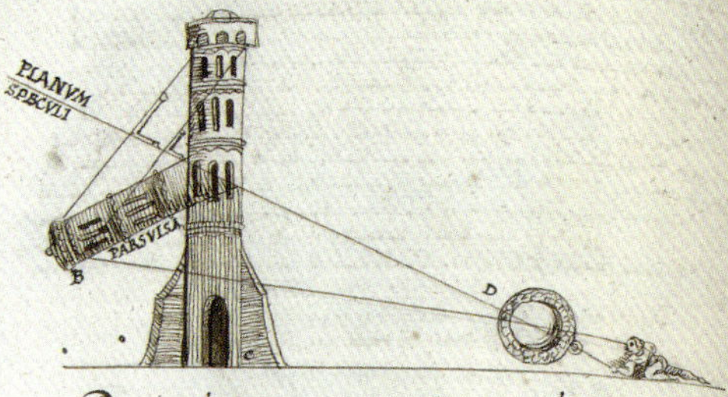

Damit du weiter den spiegel zu der messung brauchen mögest, merk
auf dis wort, wann du also für den spiegel stehst, und dazu und dar
gen magst, bis du den knopf, oder sonst ein aug des thurns im spi
siehst, müsst du mit fleiß merken, ein zeichen an der erden (gleich
gleich von dem aug herab herab nach dem blei gericht.) darum
sollt du wol gerad stehen, das dein aug nit für die zehen der füß, oder
die zehen für das aug gehen, darnach miss mit einem [illegible]
maß wie weit du von demselbigen punkt des spieg spiegel gestanden
bist (darin du das zeichen des thurns gesehen hast.) Ich setz ein gleich-
nis, du seist gestanden vom spiegel 4 ellen, und von deinen füß
zu dem aug sind 3 eln, vom spiegel zu dem thurn 40 eln, wilt du die
höhe des thurns daraus finden, so setz in die regel also. Sprich 4 geben
3 was geben 40? nach der der regel findest du 30 eln, so hoch ist der thurn.

Durch dise Figur wirdt es besser erclärt

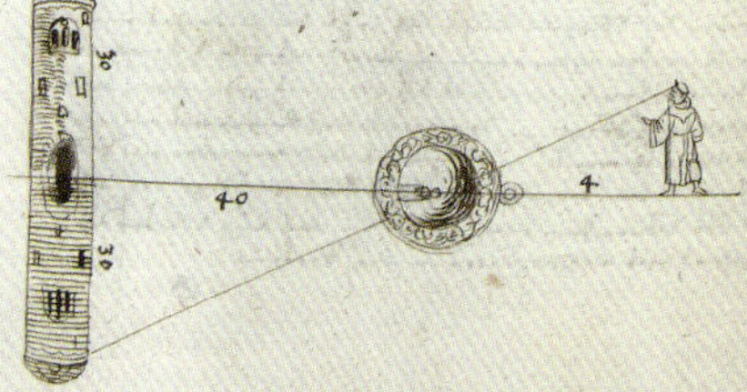

Das xv. Cap. zu m
Thurns/ wann di
Thurns nic

WAnn du zu dem grundt de
bew/ gräben/ wasser/ oder ande
hab vleyß das du von der ersten
vnd nit auff ein seyten/ vnd sch
vleyß die geleger des spigels/ vnd wie weyt
den zwayen obseruation oder absehen des T
oder höher ist dann der füß vom spigel/ als
7 schuch langk/ vnd in dem Ersten absehen
andern abschen stehest du 6 schuch von dem
zwüschen den zwayen legern des spigels sin
Sprich 2 geben 7 schuch (das ist dein höch
art der regel findest du 473 ½ schuch/ also h
vleyß haben/ das du alle ding/ die du messe
gleich als im Centro/ dann es brecht grosse
schest ainmal in der mitt den Thurn/ das a
richten. Darumb hab ich dir dise figur trewl

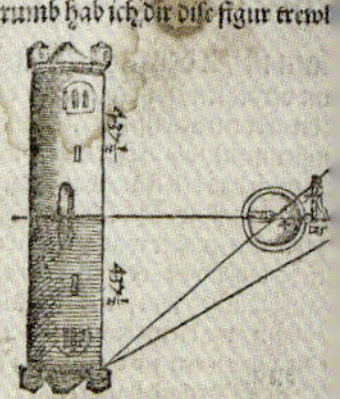

Das xvj. Capitel/ w
füß zu dem Spigel leng
sten geschicht) dann d
die höch des Turns

AM allermaysten begibt sich
ter vom Spigel stehen must dann
Exempel claer lernet. Jch setz dein f
schuch/ vnnd deine leng 7 schuch. Jn dem a
schuch/ vnnd dein leng (wie vor gesagt ist)
155 schuch/ Nun ist die frag wie hoch der th
7 schuch/ wirt in der taylung 2 ½/ Tayl auc

A Book of Instruments

Instrument Buch

Peter and Georg Apian
Ingolstadt, 1533
LE/MUN os

The simplicity of the title says it all. Peter Apian's *Instrument Buch* (1533) was the first modern general printed book about mathematical instruments. The book was aimed at a wide readership: books for scholars were written in Latin but use of German, the most widely spoken vernacular in Europe, placed it within reach of a far larger audience, which was the intention of the author-publisher. Peter Apian (1495–1552) was a Renaissance man with a vision, and a mission, to promote and disseminate practical mathematics, its techniques and instruments. With his brother Georg he set up a printing press in Ingolstadt, and the idea was not just that printing could provide books of instruction: the instruments themselves could also be printed. The quadrant, nocturnal, and so on depicted in the book double as the instruments themselves. The original purchasers could buy them on separate printed sheets and assemble their own, guided by Apian's text and illustrations; or, for a cheaper alternative, they could cut out the illustrations from the book itself.

The *Instrument Buch* was a practical and educational book, meant to be used and handled. The Museum's four copies provide the evidence that they really were: all are in a worn condition. The most interesting exists only in fragments but an early owner has neatly and painstakingly copied the missing parts from a complete example – not just the text but the illustrations too. One of the Museum's other copies is covered (bound seems too formal a word for it) in a recycled medieval parchment manuscript, creating a soft-back book that would have been pleasant to handle and easy to carry around.

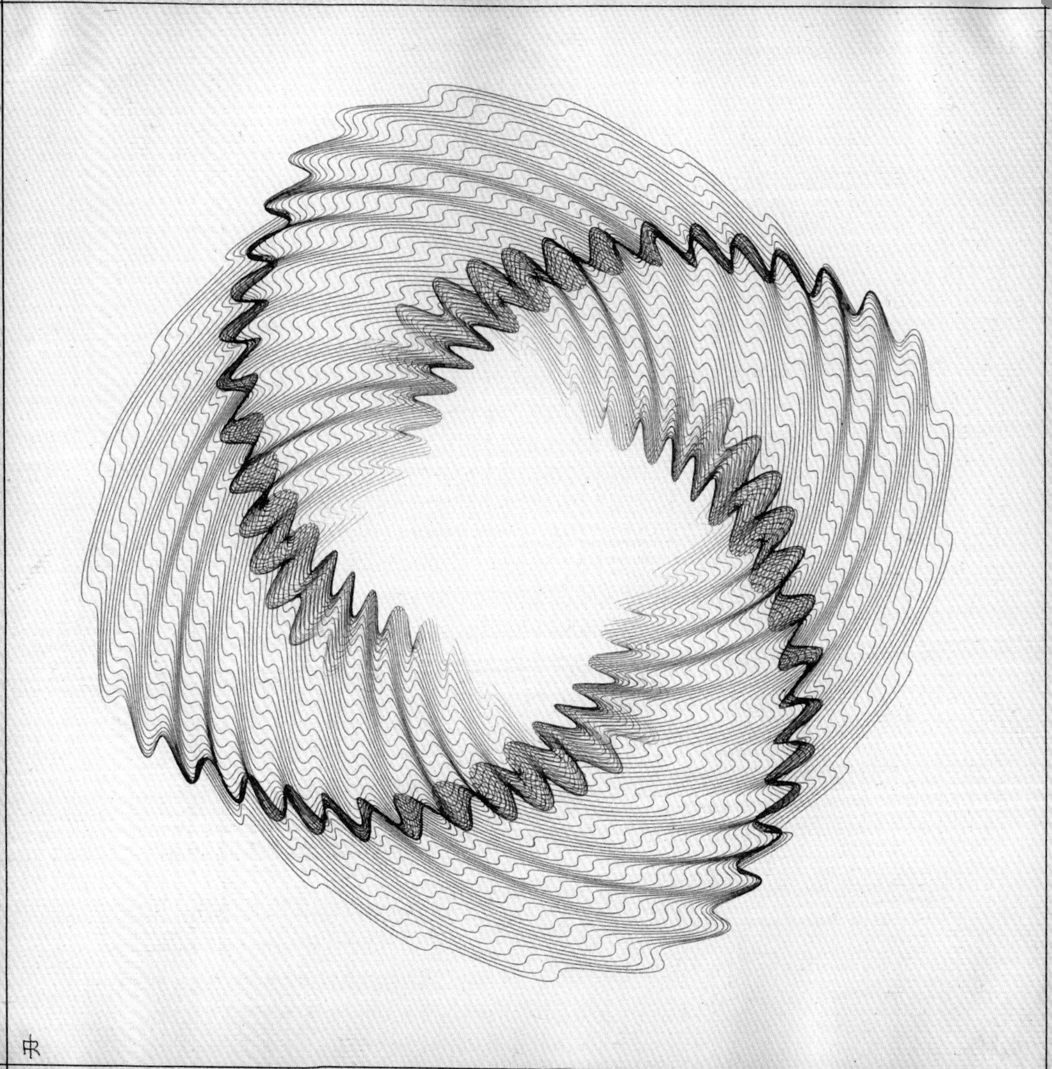

Seriously Fascinated by Curves

Album of spiraloid curves

Richard Inwards
England, 1893–8
Inv. 14732

Richard Inwards (1840–1937) was a younger member of a circle of British scientific and mechanical amateurs who were seriously fascinated by curves. Inwards happened to be a professional engineer but his friends included vicars, chemists and bookkeepers, the latter being the profession of Henry Perigal (1801–1898), the doyen of the curve enthusiasts. What especially interested them was the mechanical generation of geometric and harmonic curves, both in simple form and in the complex patterns generated by multiple repetitions. They studied the mechanics of generating the curves, the mathematics involved and the possible analogies with (for instance) sound vibrations and the orbits of celestial bodies. For a while it was a serious scientific pursuit, though during Inwards's long lifetime it came to be regarded as somewhat eccentric.

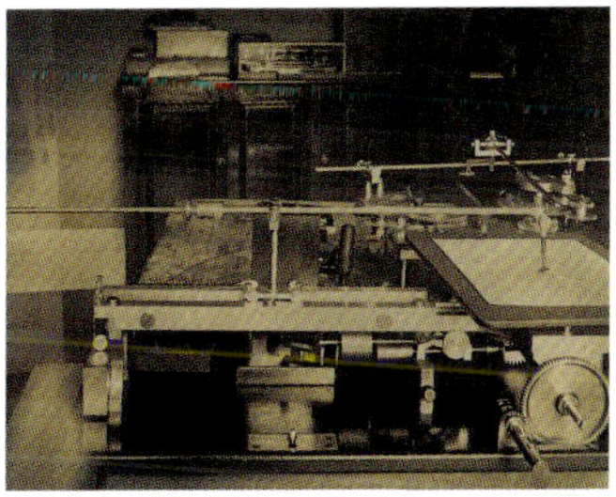

Inwards's harmonograph.

However, the visual beauty of such curves never lost its charm. Successive generations have played with popular devices for their creation, of which the 'Spirograph' is probably the best-known. In the nineteenth century, various types of harmonograph competed with other mechanical devices used to generate these stunning patterns. Inwards used a unique machine that does not appear to have survived, though we have a photograph of it. This was a geared harmonograph, based on a simpler machine made by a clergyman, S.B. Burnaby. Inwards called the results 'spiraloid curves'. In his album (which he also made and bound himself) he preserved 121 of these exquisitely beautiful designs, drawn directly by his mechanism between 1893 and 1898. If it is an unexpected thing of beauty to find among a scientific collection, what is even more remarkable is how purely it expresses the preoccupations of the Victorian amateur mechanic.

What – No Nell Gwyn?!

King Charles II's address book

England, *c.*1660

MS Gunther 60

If you wanted to consult the address book of King Charles II (1630–1685), where would you go? And once you were here – in the Museum of the History of Science, that is – who is the first person you might look for in the manuscript? Christopher Wren? Nell Gwyn, or perhaps another lady friend of the 'merry monarch'?

Unfortunately, you would be disappointed: the addresses that are recorded are not the destinations of personal letters but the correct way in which to respect protocol and titles. So the Ottoman Sultan should be addressed as follows:

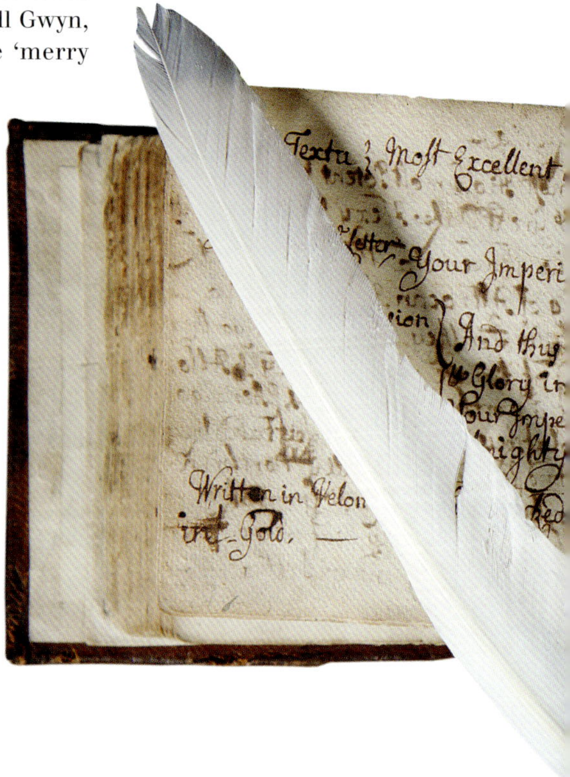

> To ye Turke
> Charles ye Second by ye Grace of ye Most High God, King of Great Brittaine, France, & Ireland; Defender of the faith &ca. To ye most High & Mighty Emperor [margin: *Sultan Mahomet] *Sultam [*sic*] Ibrahim Ham. Chiefe Lord and Commander of the Musulman Kingdome; Sole & Supreme Monarch of the Easterne Empire, sendeth greeting.
> Most high & Mighty Emperor,
> Your Most affectionate Friend C R

Written in Velome
[margin: Mr Secrys Note] Note y^{t} any Letter of business to ye Grand Signor must ever be accompanied w^{t} one to ye Grand Visier to Introduce ye matter & engage his favour.

This plain little volume, from the eclectic collection of the Museum's first curator, R.T. Gunther, is a diplomatic formulary: quite literally a pocket guide for the new King (restored to the throne in 1660 after a decade in exile). Its notes on etiquette and forms of address for foreign heads of state and dignitaries were presumably written for, rather than by, Charles. But although it is about formalities, it is itself a charmingly informal document, genuinely carried about and consulted by the King and his private secretaries.

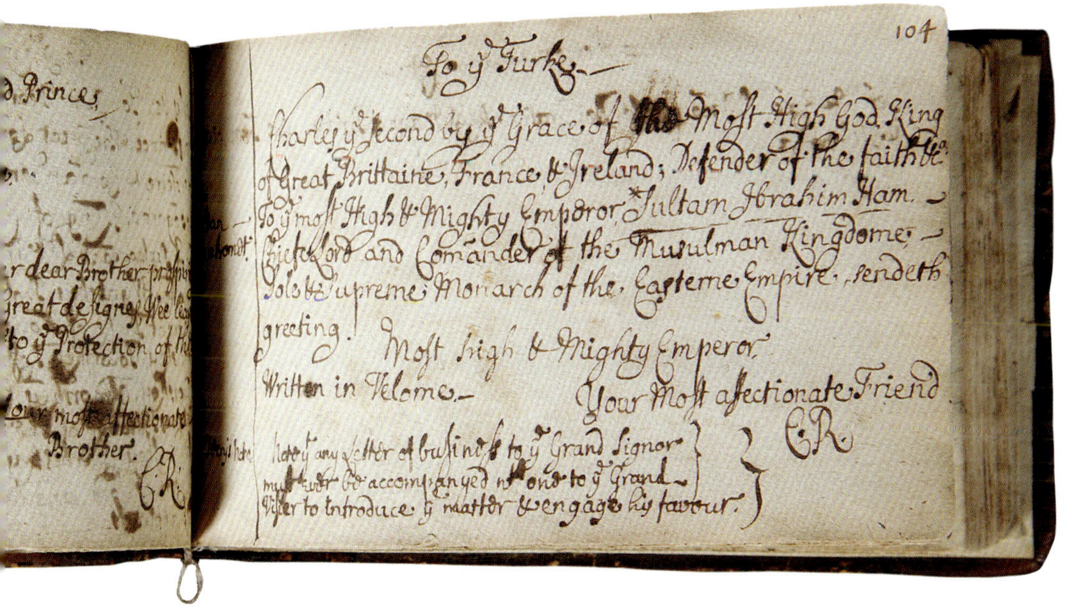

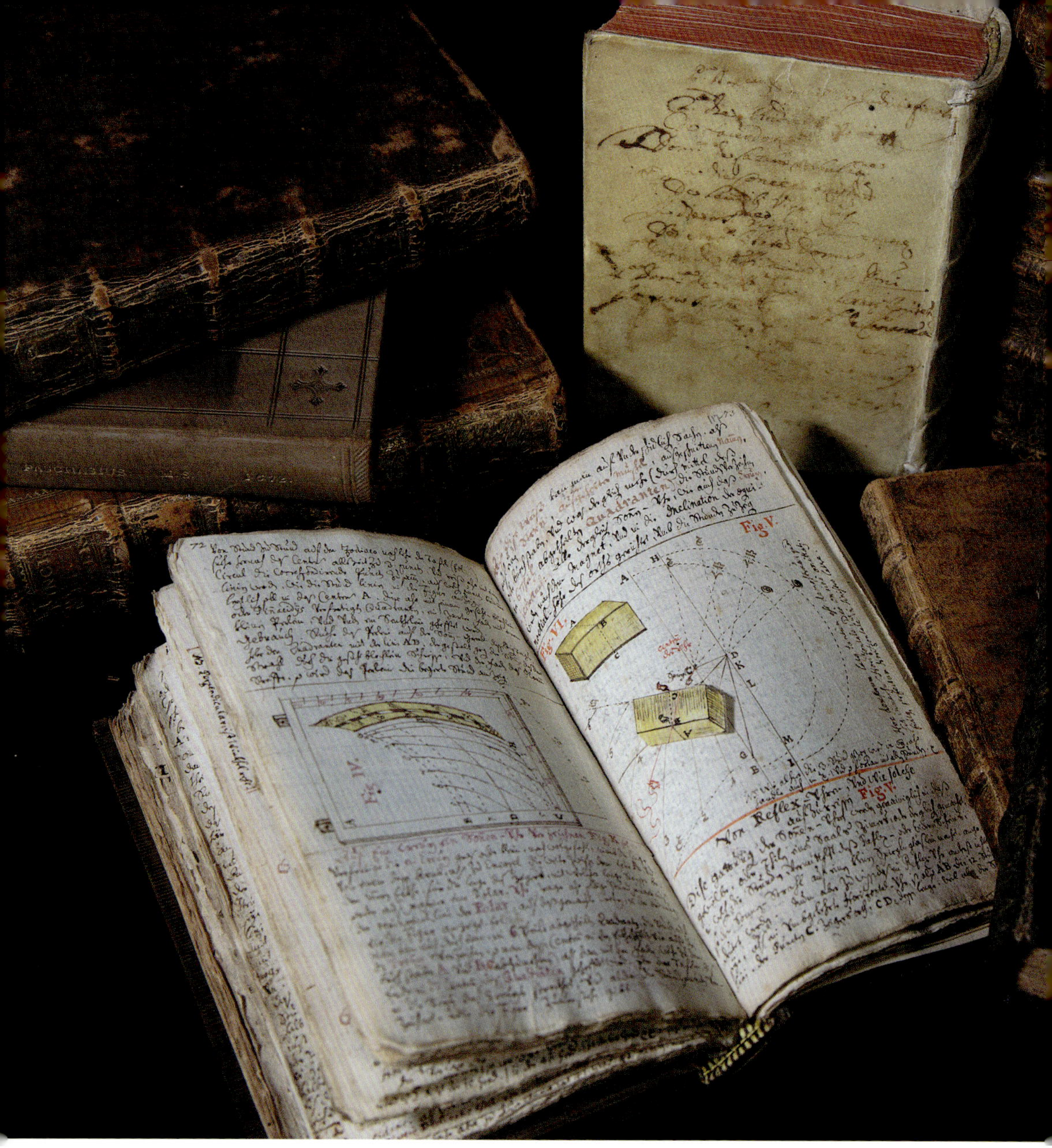

The Art of Dialling

Lewis Evans's collection of sundial manuscripts

16th–20th century

Lewis Evans, the Museum's founding benefactor, concentrated as a collector on a small range of horological instruments (as opposed to clocks). He was principally interested in sundials, both objects and texts. He acquired manuscripts as well as printed books, extending even to a small collection of printed ephemera, always focused on non-mechanical ways of telling the time.

The precisely focused nature of Evans's passion appears eloquently in his books and manuscripts. A high proportion have a neat little note written by him at the front saying which particular pages refer to sundials. Even though a work might cover surveying, navigation and armillary spheres, Evans purchased it for the sundial section.

His manuscript collection is small, about eighty volumes, yet it is a uniquely rich archive of its specialised subject. It is also very concentrated in date, chiefly from the seventeenth and eighteenth centuries. Because he collected all over Europe, it is refreshingly cosmopolitan: seven languages and at least nine (modern) countries are represented. Some volumes are scruffily scribbled notebooks or exercise books, while others are beautifully written and original monographs representing many hours, or years, of painstaking work. Most are illustrated with sundial designs, or even feature actual working instruments on paper. Some were written by students or practitioners, but inevitably many are the work of more highly educated men – teachers, priests and members of religious orders. Mathematical expertise was common among the clergy, but it is hard to resist the thought that work requiring such concentration and so many hours of patient dedication may have provided a form of religious discipline for at least some of these writers.

Father and Son

Photograph (experimental photogenic drawing) of Sir William Herschel's 40-ft telescope

John Herschel
Slough, 1839
Inv. 85037

The Herschel family was one of the most distinguished scientific dynasties of eighteenth- and nineteenth-century Britain. The German-born Sir William Herschel (1738–1822) began his career as a musician. However, it was as an amateur astronomer that he discovered Uranus, making him the first person in recorded human history to discover a new planet. His son Sir John Herschel (1792–1871) was a distinguished astronomer in his own right, but one of his many other interests was the brand-new field of photography. Inspired by the invention of his friend W.H. Fox Talbot (1800–1877), Sir John quickly perfected Talbot's process, establishing 'hypo' as the standard photographic fixative and inventing numerous alternative processes, including the cyanotype (blueprint) and chrysotype.

In 1928 Sir John's daughters Francisca Herschel and Constance Lubbock gave the Museum a handful of little paper packets containing their father's pioneering photographic experiments conducted between 1839 and 1844. This extraordinary experimental archive consists of 697 small pieces of paper, of which over three hundred are photogenic drawings or photochemical experiments. However, only four of them are images taken through a lens rather than contact prints. These were made in the autumn of 1839 and had a more personal motive: they show Sir William Herschel's huge 40-ft telescope at the family home in Slough. 30 November was the last day on which Sir John did any photography for a year, because he was busy moving house. His father's telescope was already an icon of astronomy, and the Royal Astronomical Society had adopted it as its symbol, but it was demolished within a month. The four photographs thus represent a double first: the earliest photographs of a scientific instrument and the first photographic record of a structure prior to its demolition.

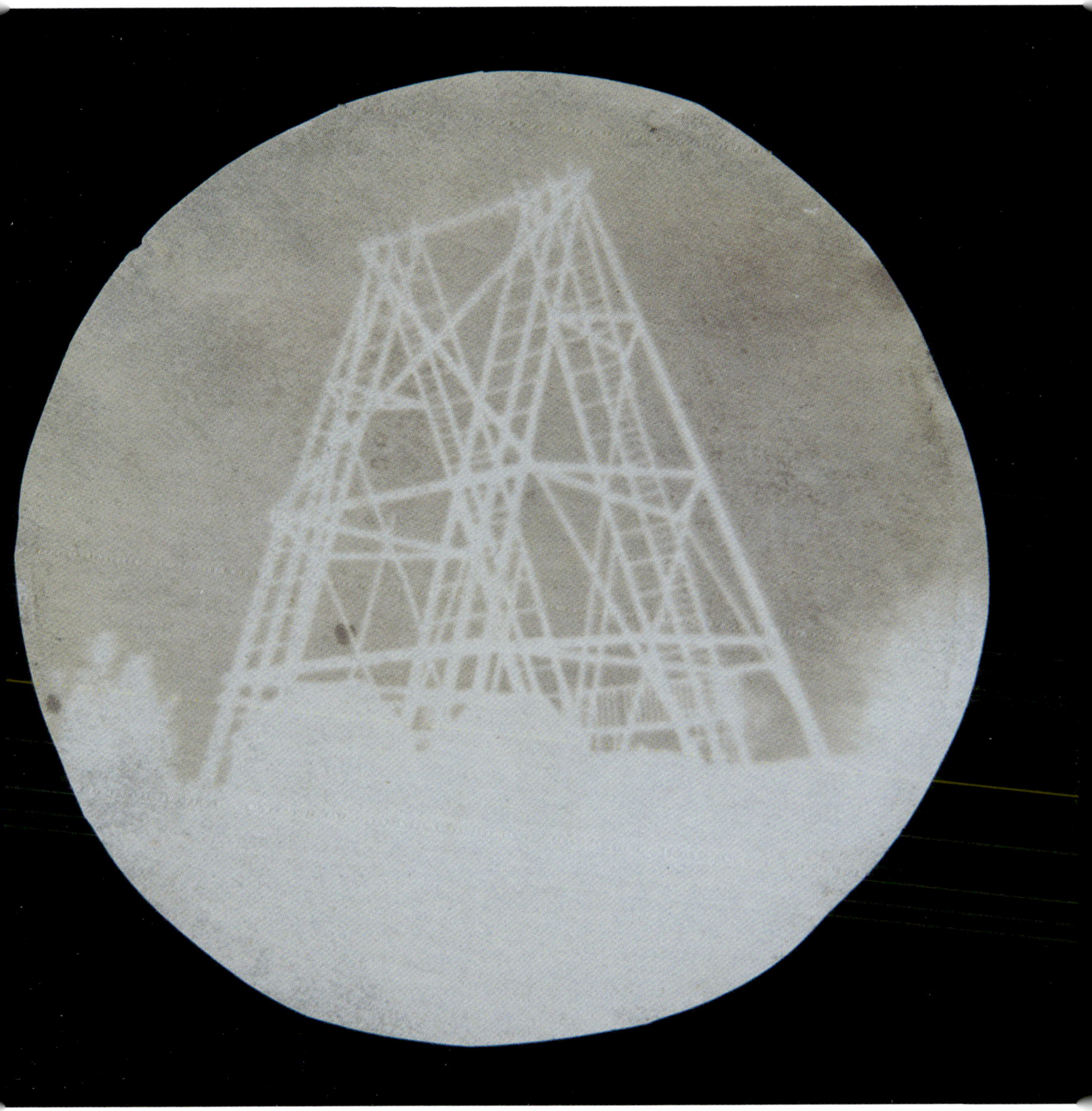

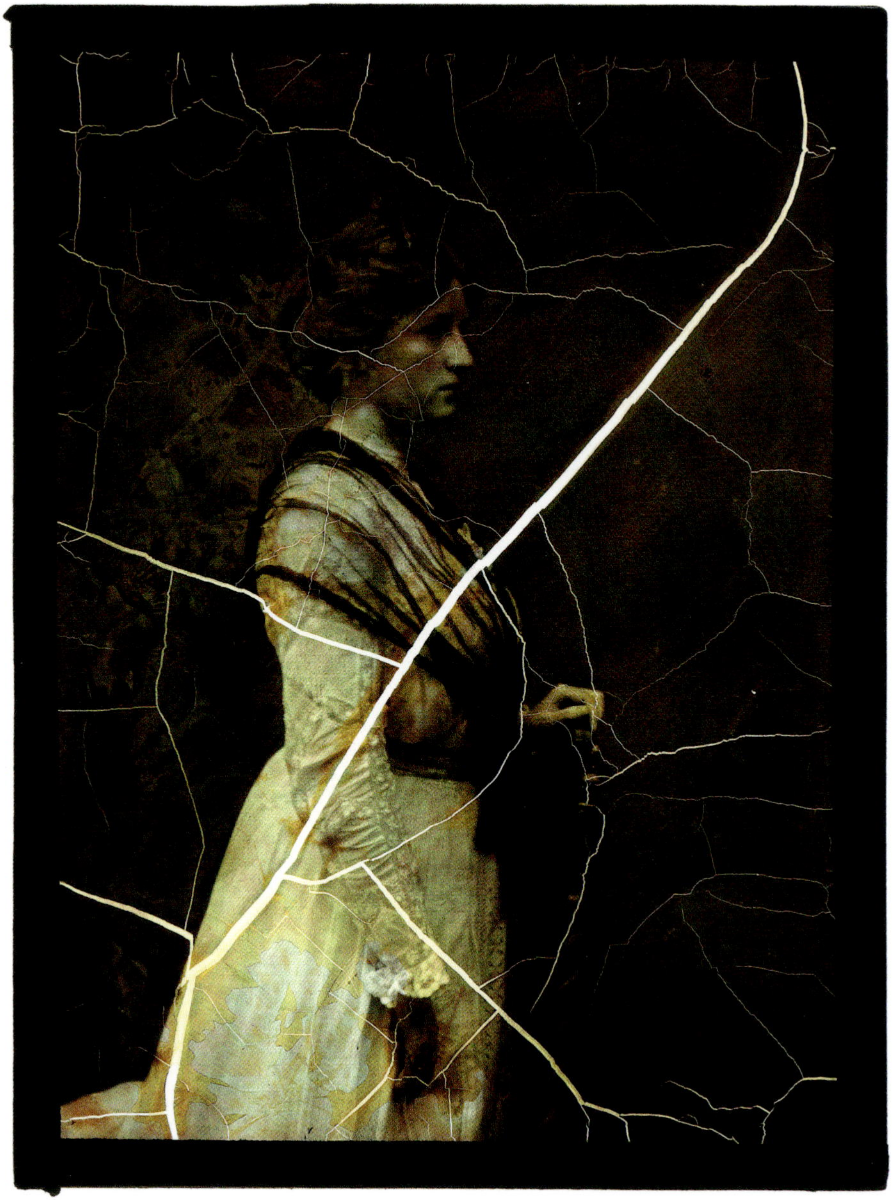

Crazy Lady

Colour photograph (Dufay dioptichrome process) of a woman in profile

William Augustus Casson
England, *c.*1910
Inv. 26048

The invention of photography was based on the darkening properties of silver salts, so only monochrome (black-and-white) photography was possible at first. Colour photography had greater technical obstacles to overcome. While its principles were understood, colour was not really feasible until the 1890s, and did not become widespread until the 1930s.

The Museum's collection covers this pioneering period. It could be represented by any number of exquisitely colourful and pretty images, especially using the most successful of the early processes, the Lumière autochrome. However, we have chosen a beautiful failure, a seriously defective early colour transparency which reveals the experimental nature of working with colour.

The photograph was taken by a civil servant named W.A. Casson, who was active around 1911 but only a handful of whose photographs have survived. His portrait of an Edwardian lady is finely posed and formally lit, its deliberate painterliness accidentally enhanced by a colour shift towards an antique sepia. This shift is a defect of the particular photographic process used by Casson: an unintended shrinkage of the emulsion caused the colours to go out of register and the whole image to crack and craze. The damage did not occur straight away, nor is it due to poor storage or the ravages of time: it is a characteristic failure of colour photographs created using the earliest form of the Dufay dioptichrome process, first introduced in 1910. Louis Dufay (1874–1936) went on to successfully reformulate the process and Dufaycolor was still in use in the 1950s.

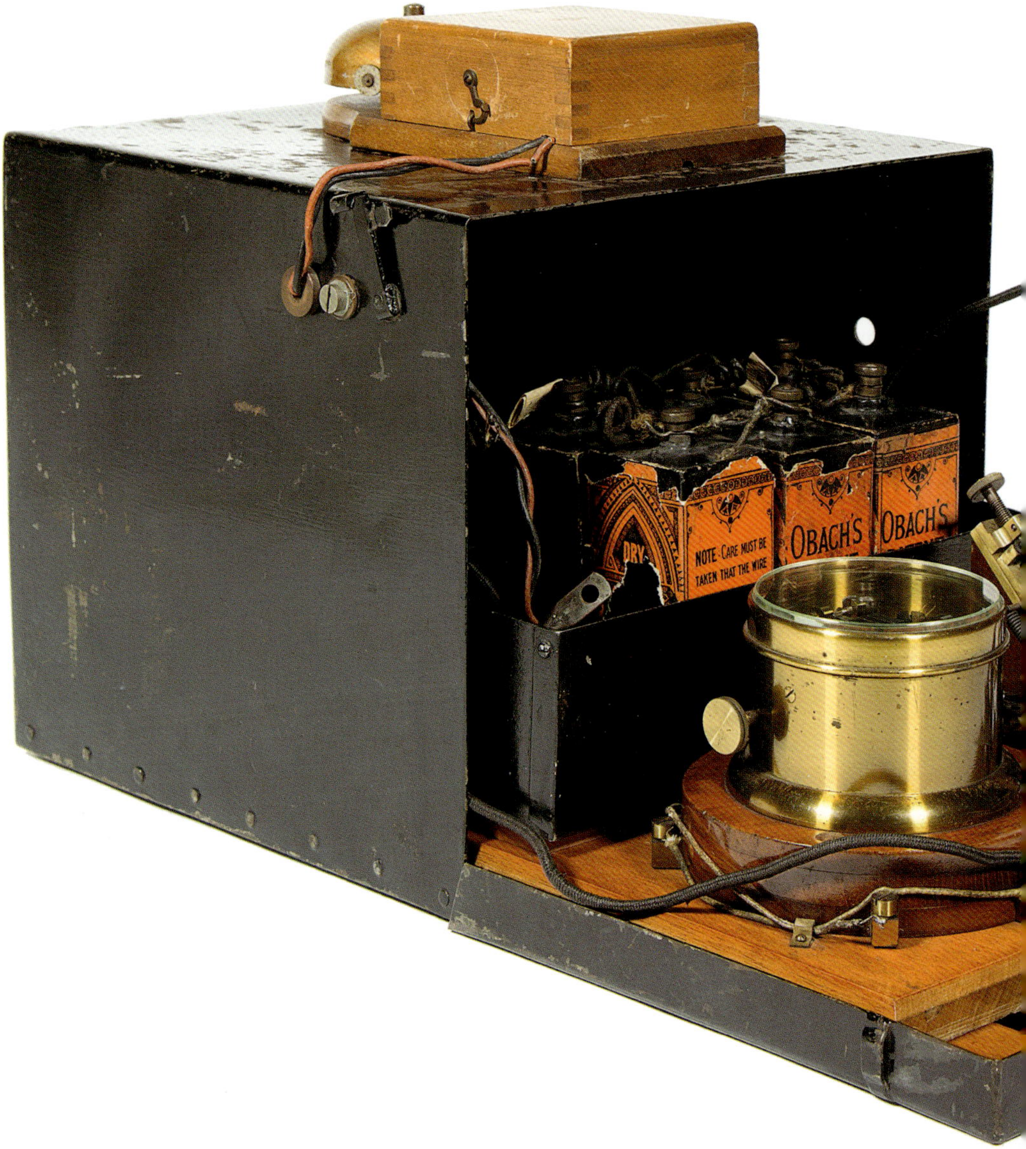
OBACH'S
OBACH'S
DRY
NOTE: CARE MUST BE
TAKEN THAT THE WIRE

Wireless World

Coherer receiver

Guglielmo Marconi
England, 1896
Inv. 86390

In December 1896 the young Italian inventor and entrepreneur Guglielmo Marconi (1874–1937) provided a demonstration for a public lecture on 'Telegraphy without Wires' held in London. Whenever a spark transmitter was operated, a bell rang on this black box, no matter where Marconi took it in the lecture room. There was no visible connection between the two and the workings of the black box were hidden (whereas we have revealed its innards).

The demonstration caused a sensation and made Marconi a celebrity. Less than a year later, in July 1897, he established the Wireless Signal and Telegraph Company (later Marconi's Wireless Telegraph Company Limited) in order to exploit his patents in the field of wireless telegraphy and to commercially market his apparatus.

Wireless communication transformed the modern world. At first it was a means of individual communication, for sending telegraphic messages in Morse code without the need for connecting cables. Early applications focused on military and maritime communications, with the value of the latter being dramatically demonstrated by the Titanic disaster in April 1912.

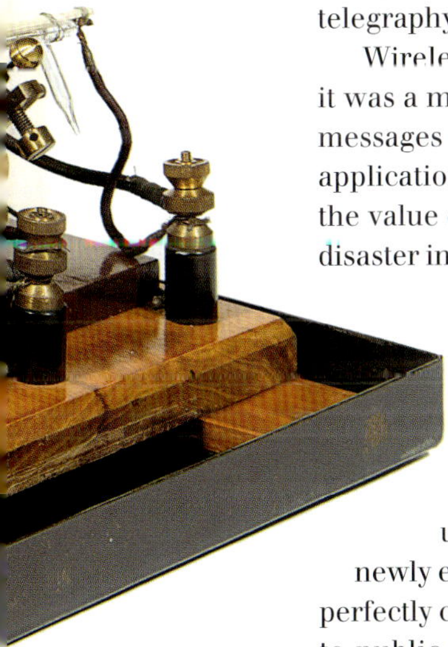

In the early 1920s radio signals began to be 'broadcast' to the many rather than the few. The first such public entertainment broadcast was by the famous singer Dame Nellie Melba from the Marconi transmitter at Chelmsford on 15 June 1920. Radio was entering the home, bringing information and entertainment, and a new industry sprang up to provide the sets that enabled a wider public to hear the newly established BBC. The commonly used expression 'listening in' perfectly captured the shift from private and individual communication to public broadcasting accessible to everyone – the origins of the wireless world we live in today.

Prize Exhibit

The Old Ashmolean Building

1683

THE MUSEUM'S BUILDING has its origin in a gift to the University of Oxford. Elias Ashmole, a seventeenth-century 'virtuoso' par excellence, offered his collection on condition that it be housed in a special 'repository'. The resulting building was created by the Oxford stonemason Thomas Wood (*c.*1643–1695). The scale and flamboyance of the deliberately out-of-proportion east portico are part of the symbolic nature of the architecture: though modest in size, its aim, like the biblical Ark, was to contain knowledge of the whole of creation.

The three-storey building featured three principal rooms. Only the top floor was for the collection. The current entrance gallery on the middle floor was a lecture and meeting room, where astronomy, experimental philosophy (physics) and geology were taught. The basement housed a chemical laboratory that served the University until 1858, as well as providing teaching space for anatomy and medicine. Here and there, pieces of ornamentation integrated the building with what it contained – elaborate locks on the outer doors and natural history carvings on the east front. Additional decoration, such as the crown on the gable and the monogram of Charles II over the north doorway, expressed both Ashmole's and Oxford's Royalist sympathies.

But it was more than just architecture. The very word 'museum' entered the English language as a result of this building: the University of Oxford did everything in Latin, and university premises were labelled in Latin over the doorways, so the translation of Ashmole's Repository was Musaeum Ashmoleanum. This first purpose-built public museum thus became famous as *the* Museum. So when we sometimes surprise people by telling them the *building* is our prize exhibit, we are not entirely joking.

Print of the Museum's east façade, 1685 (inv. 35531).

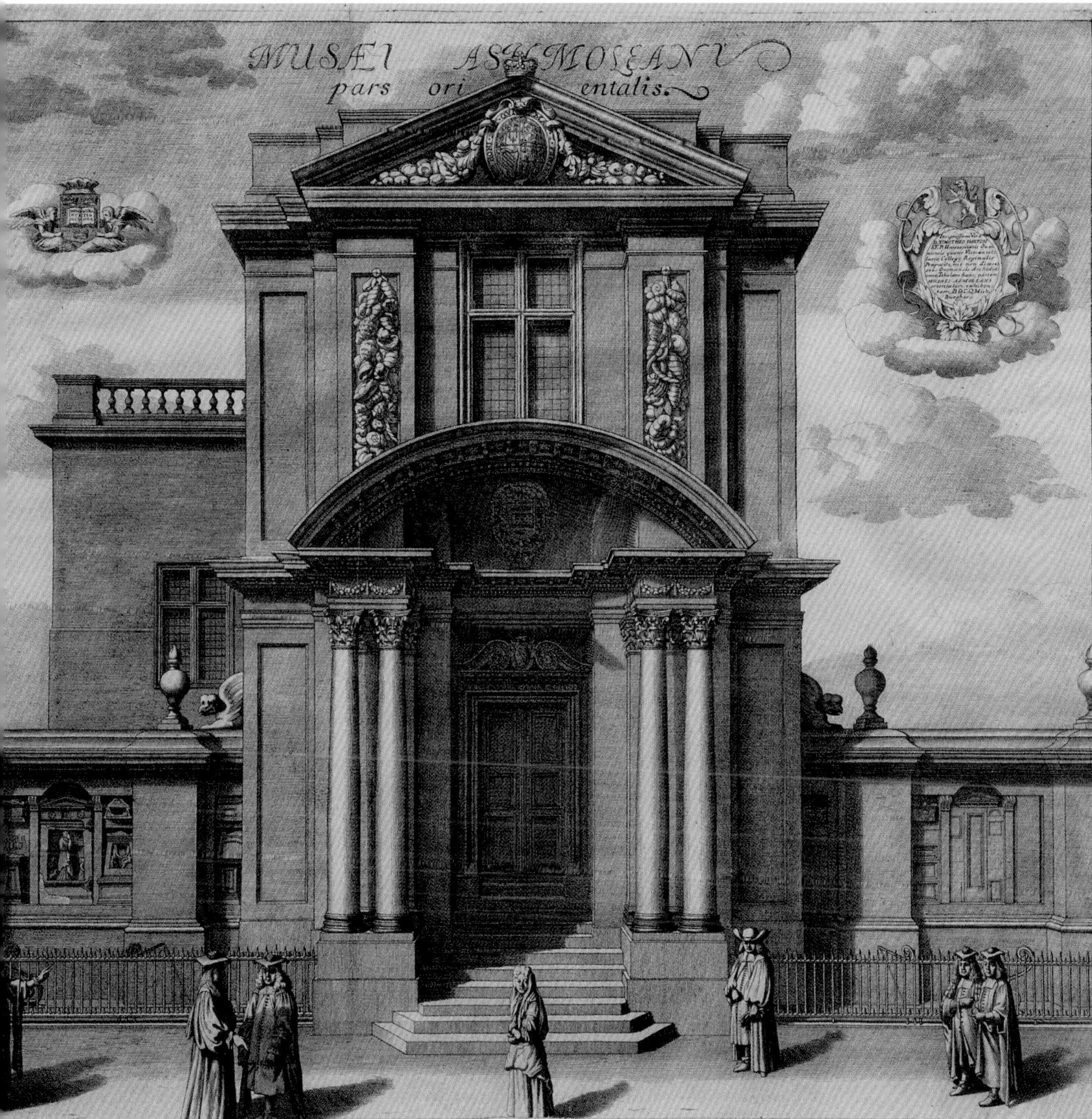
MUSÆI ASHMOLEANI
pars orientalis.

First published in 2016 by
Scala Arts & Heritage Publishers Ltd
10 Lion Yard
Tremadoc Road
London SW4 7NQ, UK
www.scalapublishers.com

In association with
Museum of the History of Science
Broad Street
Oxford OX1 3AZ
www.mhs.ox.ac.uk

ISBN 978-1-78551-031-1

Edited by Johanna Stephenson
Designed by Nigel Soper
Printed in Turkey
10 9 8 7 6 5 4 3 2 1

FRONT COVER:
Astrolabe with lunar mansions (see pp. 42–3).

FRONT COVER FLAP:
Sundial manuscripts from the Lewis Evans collection (see pp. 70–1).

FRONTISPIECE:
Main door lock.

BACK COVER:
The east façade today (see pp. 78–9).